Bill

With best regards

Nigel

Received October 1, 1990
William C. Pritchett

Vibroseis

Vibroseis

Nigel A. Anstey

PRENTICE HALL, INC.
Englewood Cliffs, New Jersey   07632

Library of Congress Cataloging-in-Publication Data

ANSTEY, NIGEL ALLISTER.
  Vibroseis / N.A. Anstey.
    p. cm.
  Includes bibliographical references.
  ISBN 0-13-957861-7
  1. Vibroseis. I. Title
TN269.88.A57 1991
622' .18282—dc20

Editorial/production supervision
  and interior design: LAURA A. HUBER
Cover design: BRUCE KENSELAAR
Manufacturing buyer: KELLY BEHR

Vibroseis is a registered trademark
of Continental Oil Company.

Printed in the United States of America

10 9 8 7 6 5 4 3 2 1

ISBN 0-13-957861-7

PRENTICE-HALL INTERNATIONAL (UK) LIMITED, *London*
PRENTICE-HALL OF AUSTRALIA PTY. LIMITED, *Sydney*
PRENTICE-HALL CANADA INC., *Toronto*
PRENTICE-HALL HISPANOAMERICANA, S.A., *Mexico*
PRENTICE-HALL OF INDIA PRIVATE LIMITED, *New Delhi*
PRENTICE-HALL OF JAPAN, INC., *Tokyo*
SIMON & SCHUSTER ASIA PTE. LTD., *Singapore*
EDITORA PRENTICE-HALL DO BRASIL, LTDA., *Rio de Janeiro*

To
John Crawford and Bill Doty,
who made
the creative leap

Contents

Preface

Vibroseis in now the dominant method of exploring for oil on land. This book seeks to explain both its theoretical basis and its practical expression in a graphical manner. As such, the book is addressed to:

- Those concerned with directing the exploration effort
- The geophysicist and technician who must implement the method in the field
- The geophysical processor, who must understand the method before the data can be processed correctly
- The seismic interpreter, who must provide the connection to geology
- The student who needs to know where Vibroseis fits in the total scheme of geophysical exploration
- Those concerned with fundamental studies of the earth's crust

Some of the material in the book is compiled and distilled from two manuals (GP308, 311) of the Video Library for Exploration and Production Specialists, written by the present author and published for limited circulation by International Human Resources Development Corporation of Boston. IHRDC's co-operation in making this material available is gratefully acknowledged. Further background material on Vibroseis, and on the seismic method as a whole, can be found in the other modules of the Video Library.

Nigel A. Anstey

CHAPTER 1

Technical
and Historical
Background

Vibroseis is a specific variant of the **seismic method,** whose major function is to determine the nature and configuration of rock layers deep in the earth.

Although Vibroseis has been used for some very interesting academic exercises—including study of the way in which the continents bump and grind—its major application is in the petroleum industry, in the search for oil and gas.

Oil and gas are known to accumulate in several situations, of which that shown in Fig. 1-1 is typical. The organic matter whose decomposition forms the oil and gas is contained in the **source rock;** under the influence of pressure and temperature, the resulting oil and gas seep up through the permeable silty rock into the **reservoir rock,** where they are trapped by an impermeable **cap rock.** The upper diagram in Fig. 1-2 shows the oil and gas trapped in a simple dome structure; the lower diagram and Fig. 1-3 show other configurations of the rock layers that can act to form a trap.

Therefore, a major part of the search for oil and gas is the search for suitable **structures** in which they may be trapped. Such structures may be 100 square kilometers in extent, or only 1; they may be 10 kilometers deep, or again only 1.

To delineate such structures as suitable for drilling, we seek to measure the depth to the reservoir at many points along a **line,** as shown in Fig. 1-4. The fluctuations in the depth measurement then define the ups and downs of the structures below some **datum** such as sea level.

Figure 1-5 is a plan view of the line (line 1), with the depth values marked

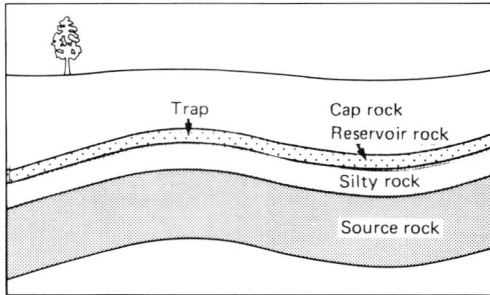

Fig. 1-1   Essential features for a hydrocarbon accumulation.

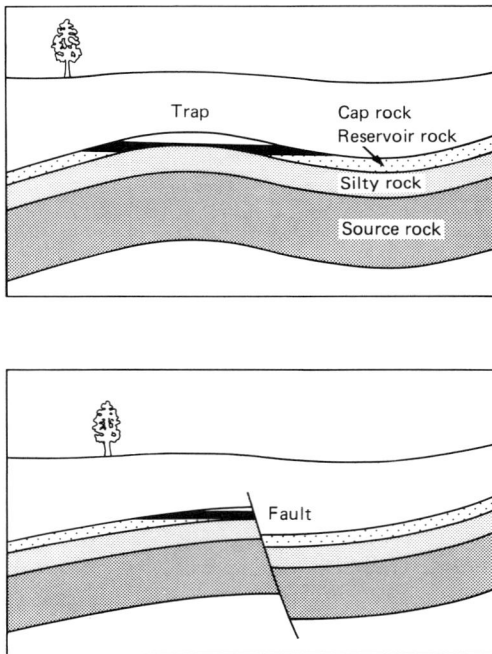

Fig. 1-2   Two important traps for oil and gas: the simple dome structure, and the upthrown fault block.

along it. Also shown is a **grid** of other lines along which depth measurements are made; the depth values from line 2 are included. By repeating the depth measurements along other lines, we can build up the map of Fig. 1-6; then we can draw **contours of equal depth,** by connecting appropriate depth values as shown. The contours define the domed shape of the structure and the best place to drill.

The essential problem, then, is to measure the depth to a particular

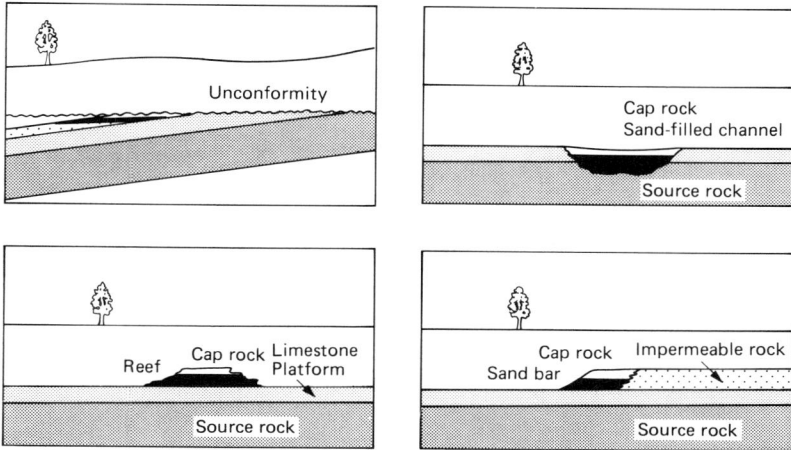

Fig. 1-3 Other trapping mechanisms: from the top, the unconformity, the reef, the sand-filled channel, and the sand bar.

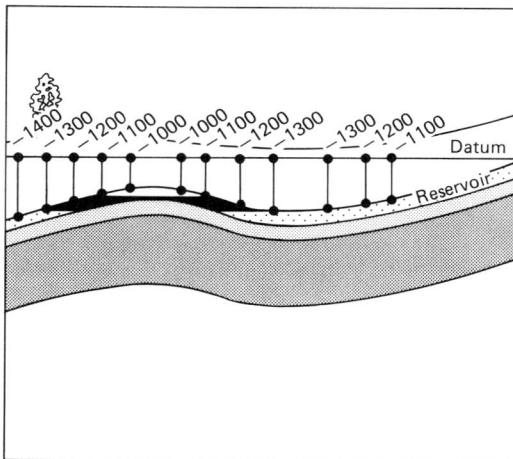

Fig. 1-4 To delineate a structural trap, we seek to measure the depth of the reservoir below a given datum (such as sea level) at intervals along a line of section.

reservoir rock at each of many measurement points along a line. It is not an easy task; our eyes cannot see into the earth, and drilling is far too expensive to be used as a mere depth-measuring tool.

The solution involves the use of **sound waves.** We normally think of sound as propagating in air, but sound waves also propagate surprisingly well in the deep rocks of the earth; in this context they become **seismic** waves.

The concept of a seismic depth measurement therefore involves making some sort of a bang at or near the surface of the ground, and timing the echoes

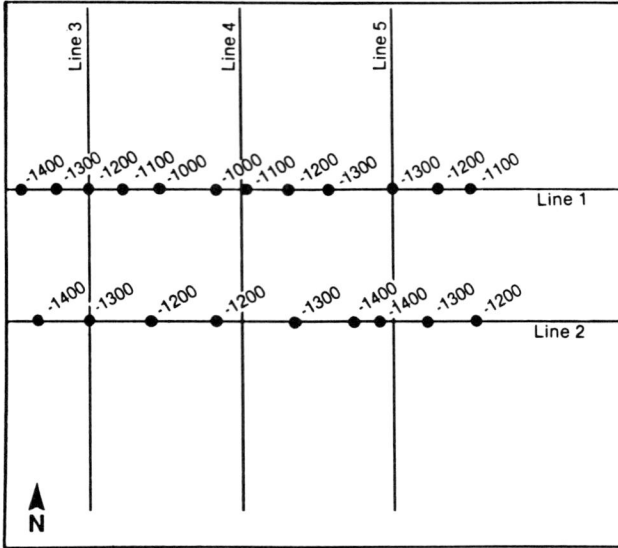

Fig. 1-5 The depth values from Fig. 1-4 are transferred to a plot of the line (line 1) on a map; then we proceed similarly on the next parallel line (line 2) of a grid of lines.

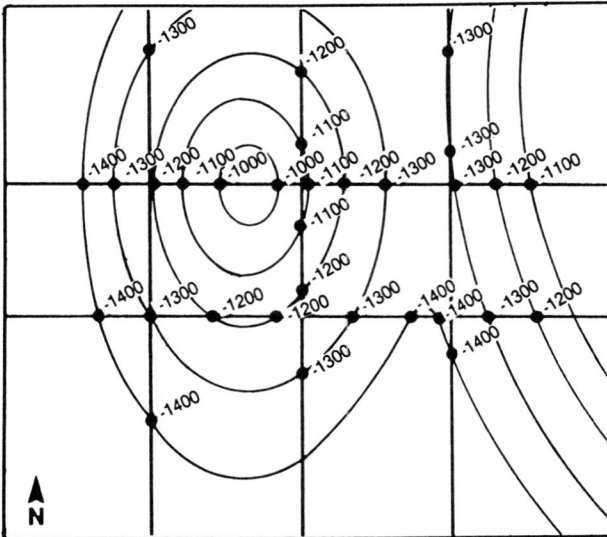

Fig. 1-6 The depth values for all lines can then be contoured by joining together points of like value.

received from each rock layer in turn. If the speed of the sound in the rocks is known (or can be determined), this echo time can be converted to a depth measurement.

Figure 1-7a illustrates the essentials of the seismic method. A source of sound waves—in the traditional practice a charge of dynamite—is activated at source point 1; the sound travels out in all directions, but one particular sound "ray" travels down to a rock reflector and back to the listening **geophone** at the surface. The reflected echo is recorded on record 1. As shown in Fig. 1-7b, the procedure is then repeated at source points 2, 3, 4, . . . , along the line.

Figure 1-7c shows a display of the results. The echo signal from the geophone is plotted downward as a function of time. The horizontal position of each such **trace** represents the position of the source-geophone observation along the line. The side-by-side arrangement of the traces, each showing its own echo, gives a display that simulates a vertical slice through the earth; this display is a **seismic section.**

Figure 1-8 is an example of such a section across a dome structure. The line is about 6 km long; it includes about 230 traces, each of which represents one source-geophone observation. The vertical scale is in seconds of reflection time; in this area 2 seconds (s) represents about 3000 meters (m) (10,000 feet, ft) of depth. Clearly, the method works.

As a seismic source, dynamite represents a vast amount of energy in a small and convenient package. In general, however, dynamite cannot be detonated on the surface of the ground; this is partly because much of its energy would be wasted upward, and partly because of the danger. Therefore, the standard practice is to drill a **shot-hole** and to load the dynamite to the bottom of this hole (Fig. 1-9).

The hole is typically 10 to 15 centimeters (cm) (4 to 6 inches, in.) in diameter and 10 to 30 m (30 to 100 ft) in depth. Figure 1-10 shows a truck-mounted drill capable of drilling to such depths in hard rock. Figure 1-11 shows a lighter unit capable of being transported by helicopter.

Figure 1-12 shows a charge of dynamite ready to go into the hole; typically, the charge might be 5 kilograms (kg) (or 10 pounds, lb) of explosive, in a cylinder 6 cm (2$^{1}/_{2}$ in.) in diameter and 1 to 3 m (3 to 10 ft) in length. Figure 1-13 shows the loading operation, by which the **shooter** tries to be sure that the charge is indeed at the bottom of the hole.

The dynamite is detonated electrically, and the instant of detonation (the **time break**) is recorded as the origin of the scale of reflection time. Often, when the charge is below the water table (as suggested in Fig. 1-9), the detonation is followed by a whoosh, as a plume of water and mud rises out of the hole (Fig. 1-14). Standard practice is to place unsuspecting visitors downwind from the hole.

Obviously, it is less trouble to plant a geophone than to drill and load a shot-hole. Therefore, instead of recording one shot into one geophone, as in Fig. 1-7, we always record each shot into a **spread** of many geophones disposed

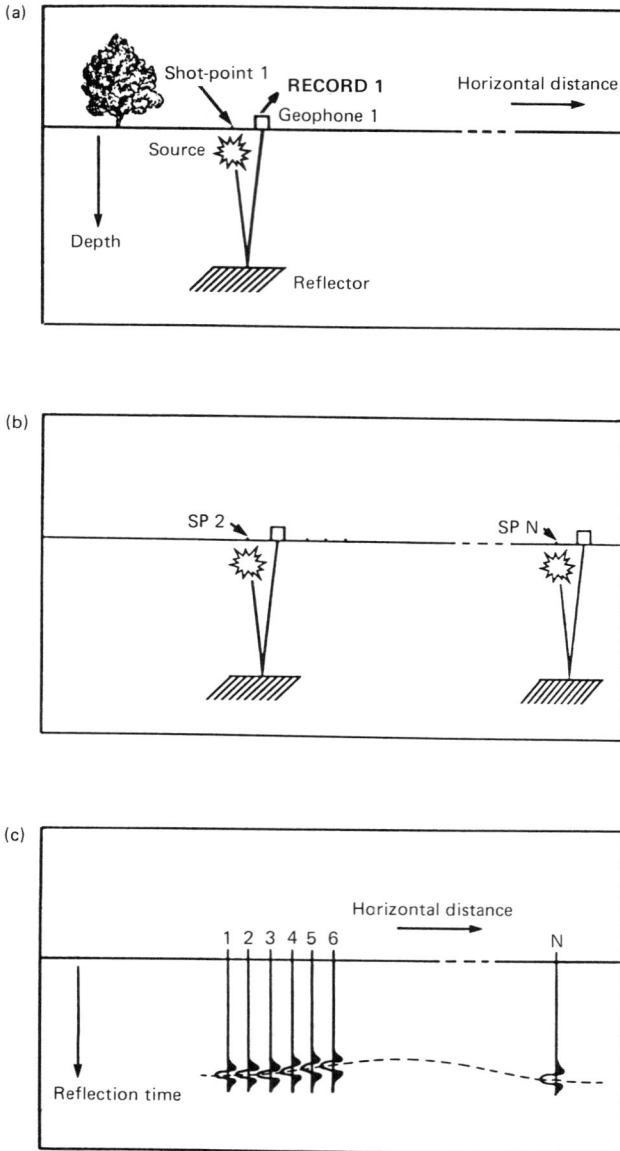

Fig. 1-7 Essential features of the seismic reflection method: the echo reflected from a source is detected at a geophone; the combination of source and geophone is moved to many positions along a line, and the results are displayed as reflection time against distance along the line.

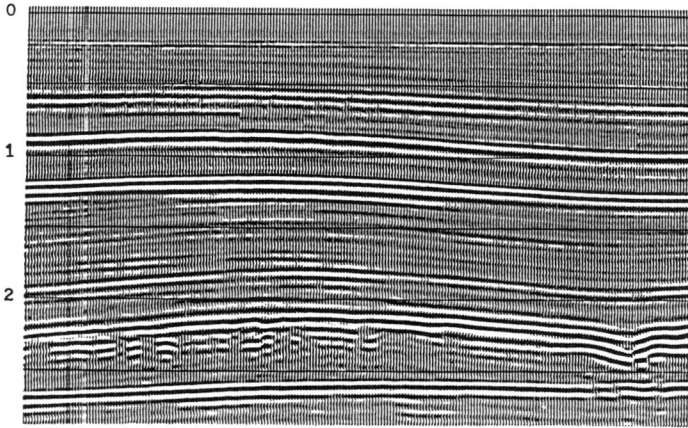

Fig. 1-8 Example of a seismic section. The horizontal dimension represents about 6 km (4 miles); the vertical dimension is 2.9 s of reflection time, which would typically represent about 4400 m (14,300 ft) of depth.

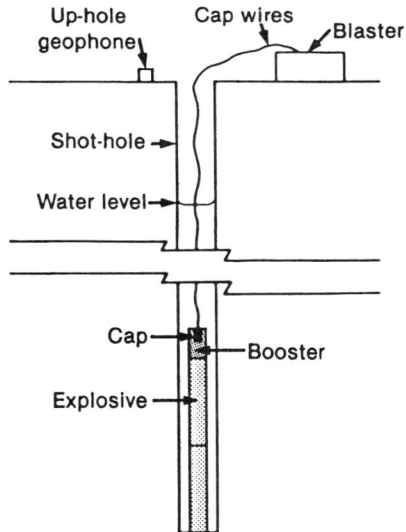

Fig. 1-9 Schematic view of the shot-hole.

along the line; this increases the extent of the deep reflector from which information is obtained with each shot. In essence, then, the traditional seismic reflection method using explosives can be sketched as in Fig. 1-15.

The distance between geophone stations, and the length of the spread, depend on the dimensions and type of the structure being investigated. Typically, the distance between geophone stations has been in the range of 35 to 100 m (110 to 330 ft), and the length of the spread—each side of the source point— has been 400 to 3000 m (1320 to 10,000 ft). In recent years the tendency has

Fig. 1-10 Truck-mounted drill (courtesy of Ardco Industries).

Fig. 1-11 Small drill capable of being transported by helicopter.

Fig. 1-12   Shooter preparing a charge.

Fig. 1-13   Loading poles push the charge to the bottom of the hole.

Fig. 1-14 Characteristic sight on a dynamite crew.

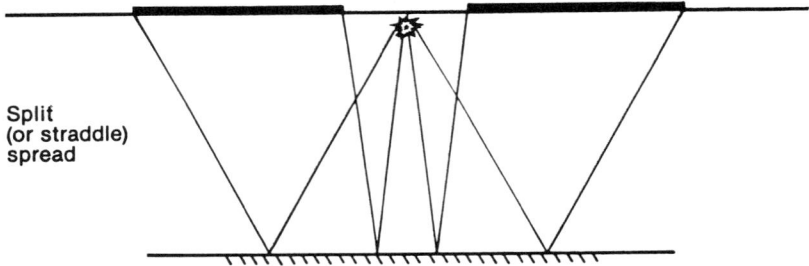

Split
(or straddle)
spread

Fig. 1-15 Basic relation between the source and the spread of geophones.

been to use a smaller distance between geophone stations, and so more geophone stations to cover the same length of spread.

As is evident from Fig. 1-16, the extent of the reflector over which reflection information is obtained is half the length of the spread. We may obtain **continuous coverage** of the reflector, therefore, by repeating the observation at distances equal to the length of the spread, as shown in Fig. 1-17. In modern practice, however, the observation is repeated at much closer spacing than this—typically 30 or 60 times within the length of the spread (Fig. 1-18). This provides **multiple coverage** of the reflector; each zone on the reflector is

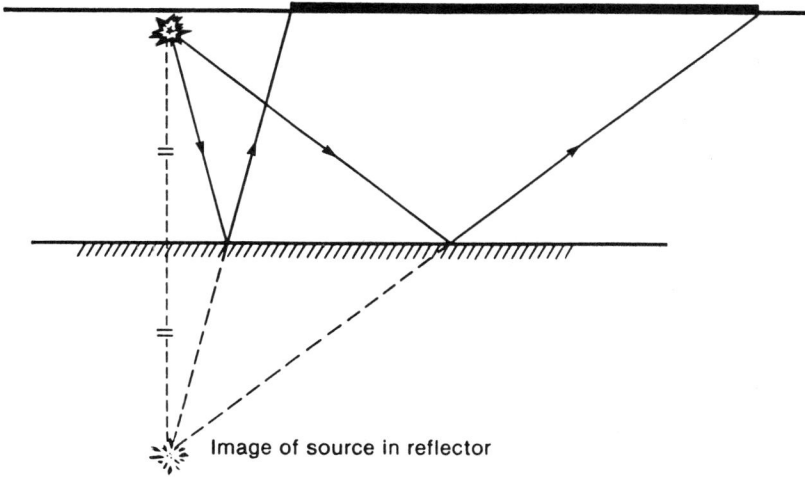

Image of source in reflector

Fig. 1-16   Extent over which reflection information is obtained is half the length of the spread.

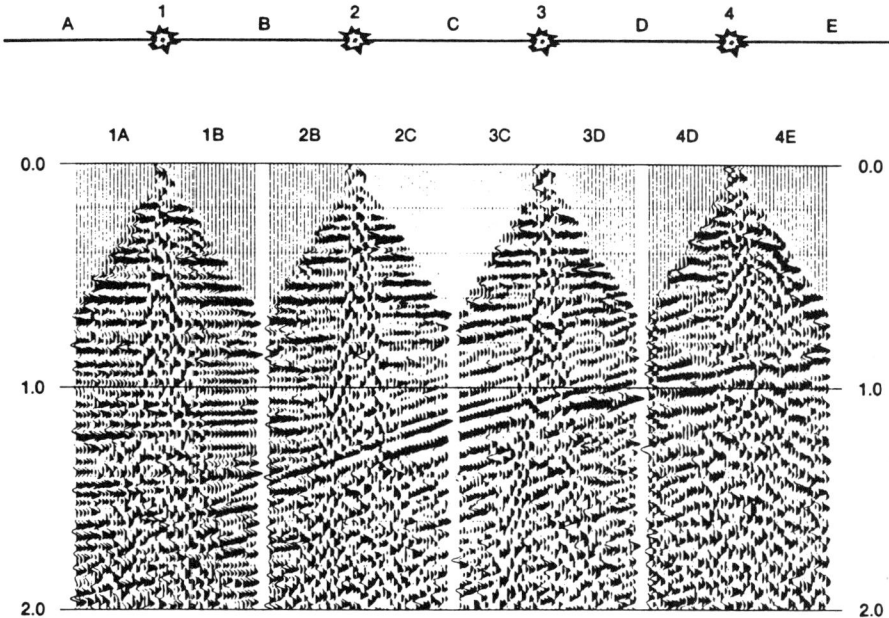

Fig. 1-17   To obtain single-fold cover, we shoot source 1 into split spread A and B to obtain the first record, source 2 into split spread B and C to obtain the second, and so on. In this way, as shown by the juxtaposed records, a continuous picture of the subsurface is obtained.

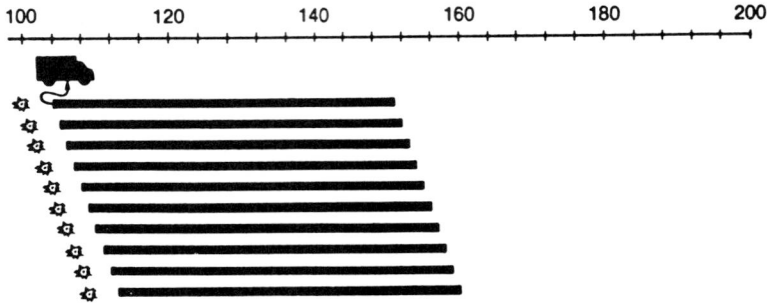

Fig. 1-18 Schematic illustration of multiple coverage. For land work, there would normally be a "back" spread as well as the "forward" spread shown in the figure, after the manner of Fig. 1-17.

sampled many times, and these many versions of the reflection information are later averaged to obtain an improved and more reliable version.

Figure 1-19 illustrates how the data obtained with multiple coverage can be manipulated to provide several or many versions of the reflection zone. From the figure it is clear that, if all the versions averaged are to represent the same reflection zone, each reflection path must be symmetrical about a **common midpoint;** this averaging of many source-to-geophone paths is therefore called **common-midpoint stacking.**

The above summary of the traditional practice of seismic work on land is obviously very condensed; fuller treatments are given in the Further Reading section. Here we now concentrate on matters concerned with the seismic source.

The first advantage of explosives is the convenience of so much energy in a very small package. A month's supply of dynamite can be stored in a small magazine, and a day's supply transported on a small truck. If a small charge is insufficient (perhaps because conditions are noisy), it is easy to double or even quadruple the charge. And the cost of the explosives is generally only a small part of the overall cost of the operation.

Another advantage of explosives is the sharp nature of the sound pulse generated in the earth. Since the essence of seismic exploration lies in the **timing** of reflections, a sharp pulse—whose time can be measured quite precisely—is much more desirable than a long lolloping rumble. It is true that the nature of the earth materials close to the charge affects the outgoing pulse and makes it less sharp; nevertheless, with dynamite we have the positive assurance that the initial input to the earth is a very sharp pulse indeed.

Another advantage of explosives springs from one of its obvious disadvantages—that the charge must generally be buried for reasons of safety. It is found that, since we have to drill a hole anyway, there is clear advantage in drilling it to an optimum depth. This optimum depth varies with the nature of the near-surface materials and with the depth of the water table; typically, as we

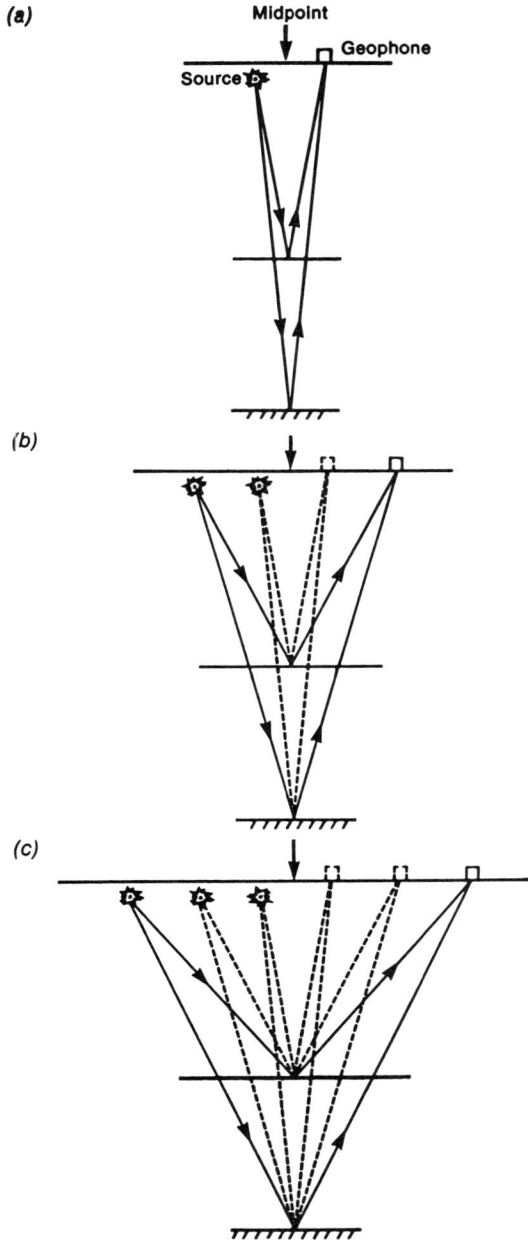

Fig. 1-19   Multiple coverage yields better results by averaging many versions of each reflection. From the data recorded with the arrangement of Fig. 1-18, we select and average those reflection paths that have a common midpoint between source and geophone.

have said, it is 10 to 30 m (30 to 100 ft). The first objective is to get below the unconsolidated (or **weathered**) materials at and near the surface; as we might expect, the sharp crack of the explosion is well maintained if we shoot in consolidated rock, but it becomes a muffled thud if we shoot in loose dry sand. The second objective is to get some preferred distance below the water table. In particular, we seek to place the charge a quarter of a "wavelength" below the water table (or, if there is no water table, below the surface); this, as sketched in Fig. 1-20, has a reinforcing effect on the downgoing pulse—like a mirror behind a lamp. It is true that this trick (which we call "using the ghost reflection") lengthens the pulse somewhat and makes it more cyclic (or "leggy"); on balance, however, the ghost reflection is almost always beneficial.

Another advantage of buried explosives is concerned with the generation of **ground roll.** As we might expect, an explosion near the surface of the ground generates not only the downward compressional pulse that we need for our reflections, but also a pattern of spreading **surface waves** (somewhat like the waves caused by dropping a stone into a pond). We often find ourselves trying to record weak reflections, from deep in the earth, against a background of these strong surface waves. Again, as we might expect, the surface waves can be made less strong by burying the explosive deeper.

In addition to surface waves, an explosive source generates another family of horizontally propagating waves, which we call **shear refractions.** Although the mechanism involves the details of the shallow layering, and hence

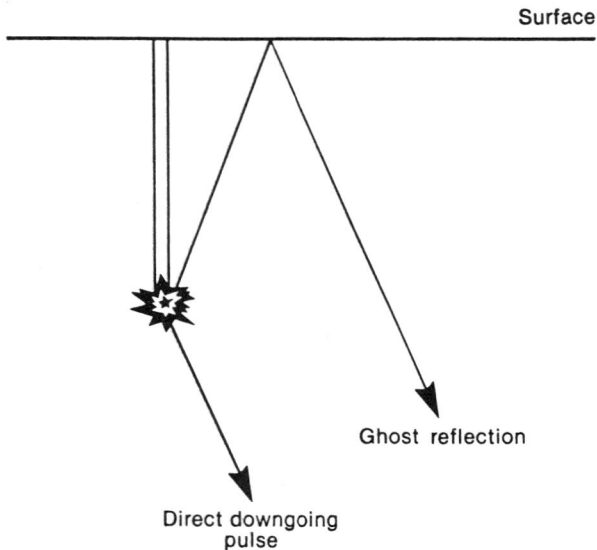

Fig. 1-20   The placing of the charge a quarter-"wavelength" below a reflector above it, and the consequent reinforcement provided by the ghost reflection.

is strongly dependent on local conditions, it often remains true that a deeper source minimizes these undesired waves also.

If an explosion "blows out" (Fig. 1-21), the bang can be heard in the air; an **air wave** propagates down the spread, additionally complicating the detection of the deep reflections. This can be prevented, of course, by deeper burial of the shot and by hard tamping.

We see, then, that surface waves, shear refractions, and the air wave (which together we call **source-generated noise**) are all harmful to our chances of recording deep reflections, and that they can all be minimized by deep burial of the explosive charge.

In many areas, the surface waves and/or shear refractions are so serious that extremely deep burial of the charge would be necessary to reduce them to an acceptable level. We need another solution. Traditionally, this involves the use of an **array** of geophones, in place of a single geophone, at each geophone station. In essence, the scheme is to lay out a group of geophones over one wavelength of the troublesome horizontally propagating waves (Fig. 1-22); then, since half the geophones are going up and half are going down, the troublesome waves are significantly attenuated. The deep reflections, arriving at the array from below, are substantially the same at all geophones; therefore, the array improves the ratio of reflected signal to source-generated noise. This approach, as we shall see, is not without its problems. However, it remains true that arrays can help to reduce the depth to which the charge must be buried, and that when the arrays are insufficient we still have the last resort of burying the charge deeper.

At this stage, the practice of using explosives as a seismic source may be sketched as in Fig. 1-23. For simplicity, the three components of the source-generated noise are shown merely as arrows.

The final advantage of buried explosives again concerns the problematic

Fig. 1-21   A shallow shot blows out, into the air.

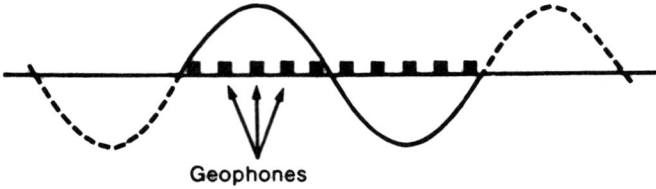

Fig. 1-22 Concept of a geophone array: for a reflection, coming up close to the vertical, all the geophones reinforce; for a surface wave having a wavelength equal to the effective length of the array (as shown), half the geophones cancel the other half.

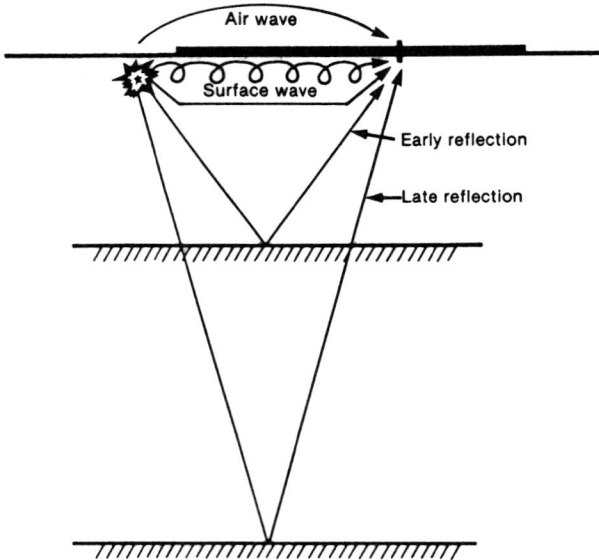

Fig. 1-23 In general, the velocities of the source-generated noise waves are less than those of the reflections, and so we may be trying to record reflections at a time dominated by the source-generated noise.

weathered layer at and near the surface. Not only is this an unsatisfactory material in which to shoot, but its natural variability—of thickness, and of the velocity with which seismic waves travel through it—means that there is great merit in shooting *below* it. Then we can measure not only the shot-to-reflector-to-surface time (the **reflection time,** observed at the geophone array), but also the direct shot-to-surface time (the **up-hole time,** observed at the up-hole geophone) (Fig. 1-24). Manipulation of these times, plus knowledge of the seismic velocities in the consolidated rocks below the weathered layer, allows us to calculate **static corrections** by which the reflection times can be corrected to

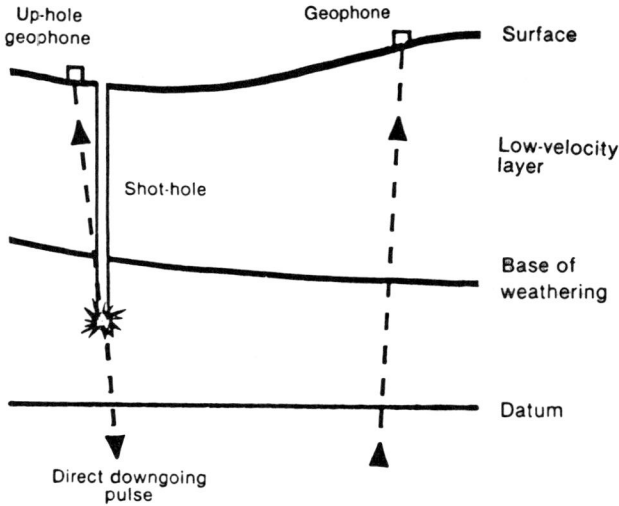

Fig. 1-24   Contribution of the up-hole time to the estimation of datum corrections.

the datum. This positive advantage arises directly from the fact that the explosive is buried below the variable weathered layer.

With all these advantages established, let us now consider the disadvantages of an explosive source.

The first disadvantage, clearly, is the need to drill the hole. Although shot-hole drilling rigs are reliable and the techniques of drilling are well established, it remains true that the need for shot-holes brings many problems. In hard-rock country, where the drilling is slow, several rigs may be required to give a reasonable rate of progress; the rigs (and their drill bits) become expensive. In other areas the supply of drilling water can be a problem. Once drilled, shot-holes often collapse; this may dictate immediate loading of the charge, which then represents a sleeping hazard until the recording crew arrives (perhaps the next day) to detonate it.

Another disadvantage is the general perception of danger in the use of explosives. Although proper handling reduces the risk to negligible proportions, and accidents are extremely rare on seismic crews, it remains inescapable that explosives make the public nervous.

Another disadvantage is the risk of damage to buildings, pipelines, wells, and other construction. This risk is commonly much exaggerated, and the seismic lore contains many tales of householders claiming compensation for cracks that proved to be full of cobwebs. However, houses have indeed been damaged from time to time; this has resulted in the imposition of "minimum shooting distances" that reflect worst-case conditions, and in any inhabited area these inevitably introduce large gaps in the subsurface coverage. Certainly, we cannot even consider using explosives for the seismic delineation of a structure under a town.

Even in the country, the seismic method using explosives is effectively debarred from working along roads. This is a major disadvantage in forested and other areas where the roads represent the only easy access. In effect, the seismic line is forced to go cross-country; this inevitably involves delays and difficulties associated with hedges, fences, river crossings, valuable crops, and cattle. (Of course, it is also true that we prefer our seismic lines to be *straight,* and where the roads reel about like drunks we may be forced to go cross-country anyway. However, we would certainly prefer to *choose* whether to shoot along the roads or not; with explosives, we generally do not have the choice.) Where we damage the crops we must pay compensation. In all cases the hole must be filled in after use, to eliminate hazard to farm animals or vehicles. All this represents an unwelcome burden—of seeking permits, of finding access, of negotiating damage claims, of restoring the surface, and of sheer physical difficulty in doing the field work.

The present author joined his first seismic crew in 1948, in the mudflats and deserts of southwest Iran. In such a setting, there seemed little need to search for an alternative to dynamite as a seismic source. The lines were laid out straight across the desert; there were no problems of access, no buildings to damage, no landowners, and no crops. The reflection records clearly showed the benefits of the sharp pulse obtained from dynamite; the interpreters would sometimes be heard arguing about *one millisecond.* The spread consisted of a single geophone at each of 24 geophone stations spaced at 50 m (165 ft); there were therefore 12 geophones over 600 m (1980 ft), each side of the source point. Even with single geophones, and certainly with the four-geophone arrays coming into use at that time, the source-generated noise was reduced to an acceptable level by drilling the shot-hole to 15 m (50 ft); a single small truck-mounted rig could easily drill more holes in a day than the recording crew could shoot. A charge of 1 kg ($2\frac{1}{2}$ lb) was usually adequate; the cost of explosives was negligible. It is true that the shooter was a little mad, and that nobody is too keen on a mad shooter; however, the driller had a vast fund of good stories, and nobody would have wished to do without him. There was no great reason to change anything.

In other areas, however—and particularly in the populated regions of the Texas–Oklahoma oil patch—there were already the beginnings of a movement to find a **surface source.**

One of the first approaches was the Poulter method (now called **air shooting**). T. C. Poulter had been in the Antarctic, making seismic measurements of the thickness of the ice cap; he had no drill, and so started by using small charges of dynamite on the surface. Then he found that the results were distinctly better if the charges were raised on poles *above* the surface. Various explanations were given at the time, but we now know that this is because the air around the charge forms an oscillating "bubble," and that it is these oscillations—rather than the sharp crack of the dynamite itself—that get the seismic energy into the ground. Anyway, there was a brief period of enthusiasm for air shooting, during which many farmers and their cattle were frightened out of their wits by the blinding flash and ear-splitting crack of dynamite detonated in the air. Today the method is used only in extremely rough and rocky terrain, where the drilling is hard and access for a big rig is impossible.

Although air shooting was not generally successful, it did force geophysicists to think how they would solve the problem of the weathered layer when using a surface source. After an eloquent advocacy of the air-shooting method at a local geophysical meeting in 1950, the speaker was asked how he solved the weathering problem. "Ah," he said, "A weathering solution is like a toothbrush—you don't pass it around." Although nobody in the audience could be entirely sure of it at the time, we now know that there was no toothbrush, and no solution.

In the early 1950s the introduction of magnetic recording raised new possibilities for surface sources. It became unnecessary to release all the source energy in one large bang; smaller bangs could be used, and the corresponding records could be added (or **vertically stacked**) until the necessary total energy had been released.

One successful outcome of this was the weight-drop technique, affectionately known as the Thumper. Figure 1-25 shows a weight-drop truck, with the weight just about to hit the earth; the weight was about 2 tonnes, dropped from a height of nearly 3 m (9 ft). On suitable surfaces, the addition of the individual records from about eight drops would give results broadly comparable to dynamite in a drilled hole. On other surfaces no equivalence could ever be established, however many drops were made, because the pulse generated in the earth was distinctly less sharp ("more low-frequency") than the pulse generated by dynamite. This effect was aggravated by the impossibility of getting exactly the same pulse on each drop; it used to be said that "the Thumper can drop the weight four times a minute—once on each corner." And, of course, the weight often bounced. However, the Thumper was very successful in some areas, and in the deserts of North Africa it remained the method of choice for nearly 20 years.

As soon as it was demonstrated that a surface source could have sufficient energy to replace explosives, the other problems of the surface source had to be addressed.

It was immediately obvious, as one would expect, that the generation of ground roll is much more serious with a surface source. Four partial solutions were found to this problem. The first was to use more extensive geophone arrays; areal arrays of a hundred or more geophones became common, in a parallelogram or star pattern. The second was to introduce an **offset** between the source and the near geophone station, so that the ground roll (and the truck noise) would be somewhat attenuated before reaching the spread. The third was to move the weight-drop truck a short distance between drops, and so to

Fig. 1-25   Weight-drop source.

build up an extended **source array** (analogous in concept to the geophone array) during the vertical stacking process. And the fourth, which was evolving independently at the same time, was the common-midpoint stack.

It was in the third and fourth of these solutions that geophysicists came to realize that a surface source has a significant advantage over buried explosives merely by reason of its **mobility.** Thus, although the concept of a source array had also been applied to work with explosives, its routine use could scarcely be considered with deep and expensive shot-holes. A modest amount of work was done with patterns of many shallow holes, but the task of drilling and loading these holes remained very burdensome. Now, with a truck-mounted surface source, all that was necessary was to move the truck forward a short distance between drops; then the benefit of a source array was obtained virtually without cost.

Similarly, it became clear that the mobile source was the key to the smooth implementation of multiple coverage and the common-midpoint stack. The spread could be "rolled along" the line by disconnecting the geophone station(s) at the back end of the spread, and by connecting new stations at the front end; this could be done smoothly with an electrical **roll-along switch.** Then the mobile source, having just finished forming one source array, could quickly be in position to start forming the next; in this way the whole operation could be given a repetitive regularity that is difficult to achieve with dynamite in shot-holes. And this regularity, it was found, was a major factor in increasing the crew's production. Thumper crews, in suitable terrain, could get a high fold of multiple coverage more cheaply than dynamite crews.

However, the Thumper did need suitable terrain. It damaged the surface, and it could not work on paved roads. Basically, it was a tool for the desert. And desert conditions often imply a dry, unconsolidated surface layer, a deep water table, and hence a weathering problem. Exploration departments found they were drilling **dry holes** on false structures introduced by variations in the thickness of the weathering. Quickly, they learned to attach a **weathering crew** to the main Thumper crew; the weathering crew was equipped with a drill and explosives, and used these to determine the thickness of the weathered layer at a number of points over the prospect. Although this was an expensive and inelegant solution, it was less expensive and less inelegant than drilling dry holes.

Another surface source that had some success on suitable terrain was Dinoseis (Fig. 1-26). Essentially, this used a propane–oxygen mixture in a heavy chamber, and ignited the mixture by a spark; the bottom face of the chamber constituted a diaphragm, which was lowered into contact with the earth before operation.

In the context of this narrative, the most interesting feature of Dinoseis was the spark ignition; it meant that multiple units could be operated simultaneously, which was hardly feasible with the Thumper. Therefore, each truck could carry two units, and there could be several trucks; this in turn meant that the individual trucks could be smaller, lighter, and more suitable for cross-country work than the Thumper. The advantages of the source array were still available, of course; now the source array was formed both by the use of several trucks and by the movement of each truck between "pops." Thus, if there were three trucks, each emitting eight pops in unison, they could be disposed to cover the entire length of the source array, or each could cover one-third of the source array.

Fig. 1-26 Dinoseis source.

In returning to an explosive source, Dinoseis again provided a sharp input pulse to the earth. Of course, much of this sharpness was lost in traversing the unconsolidated surface layer; this is inevitable with any such surface source, but it did mean that some of the energy developed in the source was wasted.

On early models of the source, the chamber, which recoils upward after the pop, hit the ground again as it came down, and this generated an undesired second pulse; later models overcame this problem with a "catcher." Like the Thumper, Dinoseis was modestly successful in the areas best suited to it.

Today, the Thumper has been largely replaced by a variety of surface sources (ARIS (Atlantic Richfield Impulsive Source), Hydrapulse, Vacupulse, P-Shooter, and so on) that use a smaller weight, but accelerate it to higher velocities. In most cases, the weight strikes an anvil plate on the surface of the ground, rather than the ground itself; this makes for a more practical arrangement. Dinoseis also has been largely replaced by the land air gun. This gun maintains the chamber and the diaphragm, but fills the chamber with water; the propane–oxygen combination is replaced by the sudden release of compressed air into the water.

These newer surface sources, like the old ones, still exhibit all the advantages and disadvantages of surface sources. They all eliminate the cost and general undesirability of drilled holes and explosives. They all lend themselves to a smooth roll-along operation. They all generate worse ground roll, and they all offset this to some extent by making it easy to form a source array. They all sacrifice the sharpness of the pulse, by reason of absorption of the high frequencies in the unconsolidated weathered layer. And they all require some supplementary observation, other than reflection time, to solve the problem of large-scale variations in the thickness or velocity of the weathered layer.

However, there is another surface-source system, Vibroseis, that has done much better than all the rest. Because it uses a surface source, it shares the same advantages and disadvantages; it also adds a further disadvantage, in that it is undeniably more complicated. However, it brings two major advantages. The first, which is the main reason for its present dominance of the scene, is that *it can work along the roads.* The second, which is a good reason to expect its dominance to continue and increase, is that it can now overcome at least some of the absorption losses in the earth, and generate *a reflection pulse as sharp or sharper than that associated with dynamite.*

The basic principle of Vibroseis is readily understood, as follows. To obtain a reflection from deep in the earth, in the presence of the inevitable ambient noise at the surface, requires a certain amount of source **energy.** Energy is power times time. Dynamite (and all other impulsive sources) provide the necessary energy in *a short pulse of high power.* Vibroseis provides the energy in *a long signal of low power.* Because the long signal is unsuitable for precise timing of the reflections, an additional step is required (after the reflection has been obtained) to compress the long signal into a short pulse.

Vibroseis is therefore one of a class of **pulse-compressive** echo-ranging systems. Although we now know that the bats developed such a system a long time ago, the first human invention of pulse compression seems to be by a Swiss, G. Guanella, in 1938. Subsequently, a German, E. Hüttman came close, in 1940; both of these inventors were concerned with radar applications. In 1944, two British inventors, D. O. Sproule and A. J. Hughes, developed a pulse-compressive sonar system. Thereafter, the momentum returned to the radar engineers, particularly during the great surge of electronic research which occurred at the Massachusetts Institute of Technology (MIT) toward and just after the end of World War II. Most of this work was classified secret and dates are difficult to establish, but it is clear that creative work was done prior to 1952 by J. J. Bussgang, S. Darlington, R. H. Dicke, J. J. Faran and R. Hills, J. L. Jones and C. E. Kelly, and P. Rudwick. Not until 1959 were J. L. Stewart and E. C. Westerfield able to publish their classic paper "Theory of Active Sonar Detection," and not until 1960 were J. R. Klauder, A. C. Price, S. Darlington, and W. J. Albersheim able to publish their classic paper "Theory and Design of Chirp Radars."

The prime interest of radar engineers in pulse-compressive systems sprang from the problem they faced in trying to get both long range and good target resolution. To get good resolution, they needed a short pulse; to get long range, they needed high pulse energy. But they were already using power levels at which electrical flashover started to occur in the equipment; they needed a method that could *increase the energy without increasing the power*. The solution was the long signal, with subsequent compression to a short pulse.

The sonar engineers faced a similar problem, in that there is a pressure level from a sonar transmitter that cannot be exceeded; further increase merely cavitates the water on the face of the transmitter. Therefore, they also needed to increase the energy of their search pulse without increasing its power.

Although this was scarcely evident in 1952, the problem of a surface seismic source able to work on the roads is the same problem; we need to transmit the large packet of energy necessary to get a deep reflection, but without exceeding the modest power level that will damage the road.

The way in which this problem was solved is an object lesson in corporate research management: how to foster the creativity of researchers, how to use the patent system, how to keep a team focused on an objective, and how to obtain an economic return on it. The problem was solved in the research department of Continental Oil Company (Conoco), in Ponca City, Oklahoma.

On August 2, 1952, Bill Doty, a research geophysicist with Conoco, returned from a seminar he had attended at MIT and proceeded to report to his supervisor John Crawford. In an echo-ranging system, he said, it was not essential to transmit a sharp impulse. An equivalent result could be obtained, in principle, by transmitting a very long signal and then compressing it to a short pulse after reception. But there were bounds on the *nature* of the long transmitted signal; if the echo and its time were to be unambiguous, the transmitted signal *must not repeat itself* within a period equal to the greatest reflection time of interest. If it did, one reflector would appear to give rise to two echoes.

By the next morning, John Crawford had the solution to the problem of the signal. It should be a **swept-frequency** signal, a sinusoid whose frequency sweeps slowly from one limiting frequency to another. History has vindicated this choice; although many other nonrepetitive signals exist, the separation of low frequencies and high frequencies has proved to have uses beyond the mere removal of ambiguity. It was indeed a happy judgment.

Over the next few weeks the system took shape in the inventors' minds. Instead of exploding dynamite, they would shake the surface of the earth with a long vibratory signal of swept-frequency form. This they would receive with a normal spread of geophone arrays, and then compress to a short pulse; the result should look much like the traditional seismic record.

John Crawford took the proposal to his management and asked for funds to make a test. But it was late in the budget year; he would have to wait until January. In the meantime, he and Bill Doty were to make the tests in their minds; they were also to work with Bill Miller, of Conoco's legal department, to write and file the patents.

This period of enforced thought, without action, was probably the most creative episode in the history of seismic research. The minds of the scientists ran on, but the rigor of the patent attorney forced clear definitions and clear formulation. By January 1953, all the concepts were clear; the seismic problems were solved in the mind, the patent applications were formalized, and it was time to tackle the engineering.

Over the next eight years Conoco spent $8 million developing the system. There were times of encouragement and times of frustration. But the team—now including Steve Moxley, Frank Clynch, Graydon Brown, Frank Searcy, Ken Waters, Milford Lee, Pierre Goupillaud, Ross Pitman, Al Wakefield, and many others—kept on, and Conoco management had the vision to back them. By 1955, there was a thick stack of patents and patent applications. The stack included the patent that was to give Vibroseis the ability to generate a reflection sharper than that from dynamite; this was remarkable thinking for the time.

By late 1960, the engineering problems had been solved to the point where Vibroseis could be described as a workable system; to provide the final strength to its patent package, and to forestall imitators, Conoco decided to license the system to the industry.

The contractor selected for the first license was Seismograph Service Corporation (SSC), with its British and French subsidiaries Seismograph Service Limited (SSL) and Compagnie Francaise de Prospection Sismique (CFPS). The terms were tough: a large initial payment for the technology, and $150 per crew-day for each Vibroseis crew. The highest technical secrecy was to be maintained. All the derivative research was to be reported to and owned by Conoco. And the exclusivity of the license would be for only one year.

Oddly enough, the only one of the terms to prove burdensome, in the long run, was the enforced capitalization of the word VIBROSEIS, and the enforced asterisk announcing that the word was a registered trade mark.*

In January of 1961, the author and five colleagues came to Ponca City for the first Vibroseis course (Fig. 1-27). The novelty of the system, and the sharpness of the minds that had invented and fostered it, made this one of life's formative experiences.

In releasing the system outside their own company, Conoco was hoping that fresh thinking would improve the system still further. They could scarcely have guessed how fast this would happen. The weak point of the system, in January 1961, was the device used for compressing the long signal after reception (the "correlator"). By April 1961, SSL had solved this problem in the lab, and by July the magnetic correlator was in the field; it totally replaced the original device, and remained standard equipment until the digital revolution. The Vibroseis system had become *practical.*

When SSC's license became nonexclusive in early 1962, several other contractors joined the club. Prakla-Seismos, in particular, made a major commitment to the system. Others held back. In some cases this was because the geographical areas in which they operated did not lend themselves to Vibroseis operations. But in many other cases it was a simple "not-invented-here" reaction; there are always companies who delude themselves into thinking that they alone have all knowledge, and who cannot bring themselves to pay another company for its ideas. In these cases, the acceptance of Vibroseis was undoubtedly delayed by Conoco's arrogant asterisk.

Little by little, however, Vibroseis gained ground. Although the early results obtained by the method were seldom as crisp as those obtained with dynamite, the convenience of a surface source working on the road was a powerful consideration. By the 1970s, improved engineering of the equipment brought Vibroseis results to general parity with dynamite, and made the system much more reliable. By the early 1980s, more than half the land crews in the world were Vibroseis.

Some of these crews were beginning to use the approach that allows Vibroseis to *improve* on impulsive sources; this approach guarantees that Vibroseis will continue to consolidate its position as the leading seismic system on land.

In 1967, the Society of Exploration Geophysicists (SEG) bestowed the Reginald Fessenden Award on John Crawford, Bill Doty, and Milford Lee in special recognition of their invention and development of the Vibroseis system.

---

*This was a pain in the butt.

Fig. 1-27 From the right, Bill Doty, John Crawford, Milford Lee, and the 1961 technical team from SSC and SSL. The present author is sixth from the right.

In 1978, in further recognition of his leadership of the project, John Crawford was awarded Honorary Membership of the Society.

For Conoco, the years of development yielded a handsome economic return. For the inventors, there was a threefold return: the joy of invention, the plaudits of their peers, and the satisfaction of knowing that their creation was not only scientifically elegant but of immense use to humanity.

CHAPTER 2

The Vibroseis Concept

Let us review and expand the considerations that allow the replacement of dynamite by a surface source.

1. We can replace one sizable explosion, from dynamite in a shot-hole, by several or many smaller ones. The small explosions can be from explosive cord (Primacord or Geoflex) buried a small distance below the surface, or from a shot gun (Betsy), or from a gas gun (Dinoseis) or an air gun, coupled to the surface through a diaphragm. For each small shot we take a record, and then we add these records with their time breaks aligned; this, we remember, is **vertical stacking.** The signal adds linearly with the number of records, while the ambient noise adds as the square root of the number of records.

> **In principle, we can obtain the same signal-to-ambient-noise ratio with small charges, vertically stacked, as with a single large charge; then the small charges may allow operation at the surface.**

2. We can use a band-limited source. The argument here is that a dynamite explosion contains many frequencies that we do not use; if we emit only the frequencies that do us some good, we need less of a bang. For example, let us suppose that the signal from a buried dynamite explosion is similar to a large spike (Fig. 2-1a). Then we remember[1] that the amplitude spectrum of a spike

---

[1]The reader requiring a nonmathematical introduction to the concepts of amplitude spectra, phase spectra, and linear systems is referred to GP 201 *Wiggles* of the E&P Library (IHRDC,

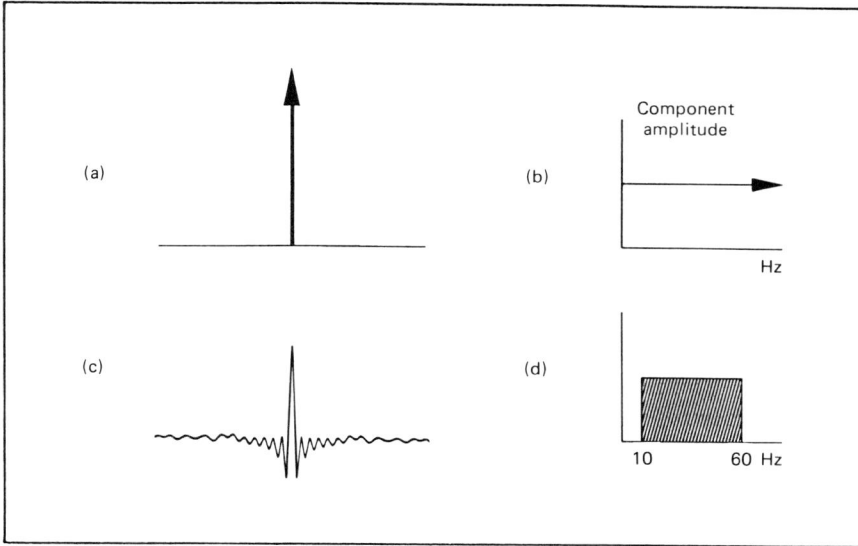

Fig. 2-1   Representation, in time and in frequency, of spike and band-limited signals.

contains all frequencies equally (Fig. 2-1b) and that the phase spectrum is zero. Let us suppose that by analyzing the resulting record we can be quite sure that the reflections have no significant frequency content below 10 hertz (Hz) or above 60 Hz. So in place of the spike we could use the signal of Fig. 2-1c, the zero-phase signal whose amplitude spectrum is flat from 10 to 60 Hz. Then, despite the fact that the peak amplitude of Fig. 2-1c is much less (and therefore much safer) than that of Fig. 2-1a, the reflection record we obtain is *identical*.

There is no point in using energy to generate frequencies that do us no good.

3. The third consideration recognizes that, for a certain level of ambient noise, a certain source *energy* is necessary if we are to detect a reflection at a certain depth; then it invokes the fact that energy is power times time.

A certain source energy can be obtained with a very high power for a very short time (as in a spike) or with a lower power for a longer time.

The Vibroseis method uses all three of these approaches. But it is the third, the use of a long signal of modest power, that is the essential feature of

---

Boston), or for a less complete account to Chapter 2 of *Seismic Prospecting Instruments* (Gebrüder Borntraeger, Stuttgart), both by the present author.

Vibroseis. And it is the modest power that really makes practical the use of a surface source. No one would stand by a detonating explosive charge on the surface, but a Vibroseis source, emitting the same *useful energy,* does no more than tingle the toes.

Now we must ask how it is possible to obtain explosion-type records using an emitted signal that is long. In Fig. 2-2 we visualize a black box, whose only effect is that *it delays some frequencies more than others.* Let us say, loosely, that it imposes a short delay on the low frequencies, an intermediate delay on the intermediate frequencies, and a long delay on the high frequencies. Into it we inject the symmetrical band-limited signal of Fig. 2-1c; this, we remember, contains equal amounts of all frequencies from 10 to 60 Hz. What happens? The black box passes all these frequencies equally, but with different delays; the output is a constant-amplitude signal whose apparent frequency sweeps progressively from low to high. This swept-frequency quasi-sinusoid is the signal we shall call a **sweep.**

In Fig. 2-3 we take the output of the black box—the sweep—and we inject it into a white box. This is just like the black box, except that is imposes a long delay on the low frequencies and a short delay on the high frequencies. In fact, the delay–frequency relation of the white box is exactly *complementary* to that of the black box. Now what comes out? Clearly, it is the original signal, the one we injected into the black box. It must be so, because neither box has affected the amplitude spectrum, and because at every frequency the sum of the two delays is constant.

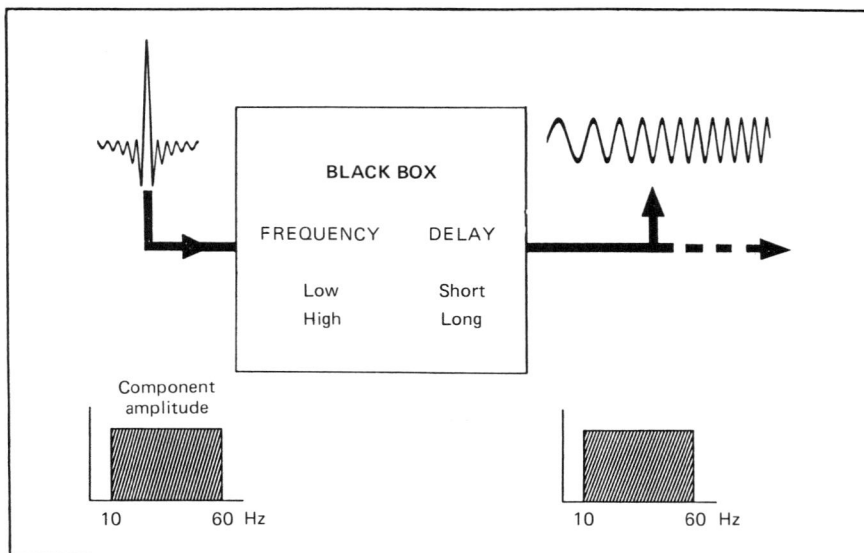

Fig. 2-2 Formation of a long sweep from a short pulse by the application of frequency-dependent delays.

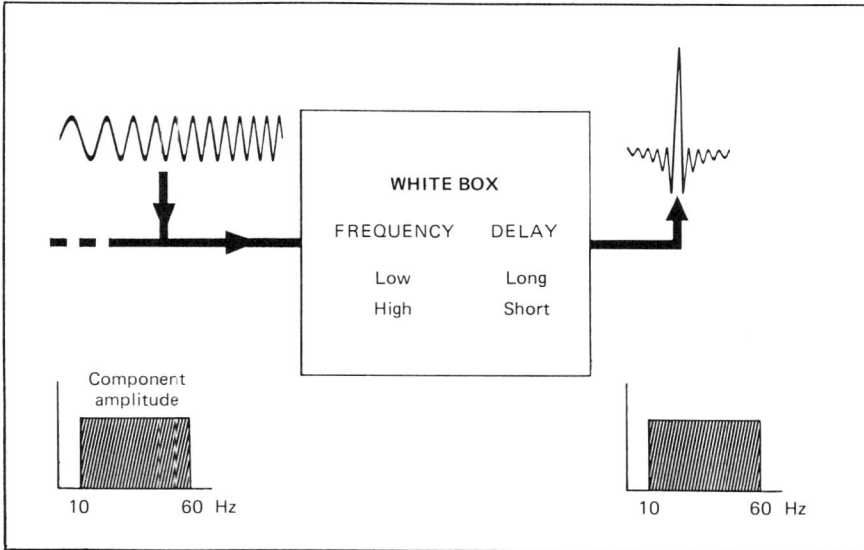

Fig. 2-3 Formation of a short pulse from a sweep by the application of complementary frequency-dependent delays.

In Fourier terms, we can think of this process as in Fig. 2-4. Very loosely, the frequencies that are in phase at the center of the initial input signal (a) are *strung out sequentially* in the sweep (b). Then they are brought back into phase in the final output (c), which is therefore the same as the initial input (a).

So the *black* box acts to expand a short pulse of high peak amplitude into a long sweep of low peak amplitude. The *white* box acts to compress a long sweep of low peak amplitude into a short pulse of high amplitude. And the energy in both forms of the signal is the same.

Time and again, in the commercial literature, we find Vibroseis called a low-energy system. This is wrong. All systems of comparable penetration must employ comparable useful energy.

Vibroseis is not a low-energy system; it is a low-power system.

Now for the big step. In Fig. 2-5 we preserve, at the top, the simple sequence of a black box and a white box, generating a sweep and then compressing it back to a symmetrical pulse. But we also route the sweep to a vibrator and inject it into the earth. Some time later, the sweep is reflected back to the surface and detected by a geophone. Then we pass the output of the geophone through a second white box, where it also is compressed back to a simple symmetrical pulse. (We ignore, for the moment, frequency-selective effects in the earth.) So we have a direct pulse from the first white box, and a reflected (and delayed) pulse from the second white box. The time T between them is, of course, the travel time of the reflection.

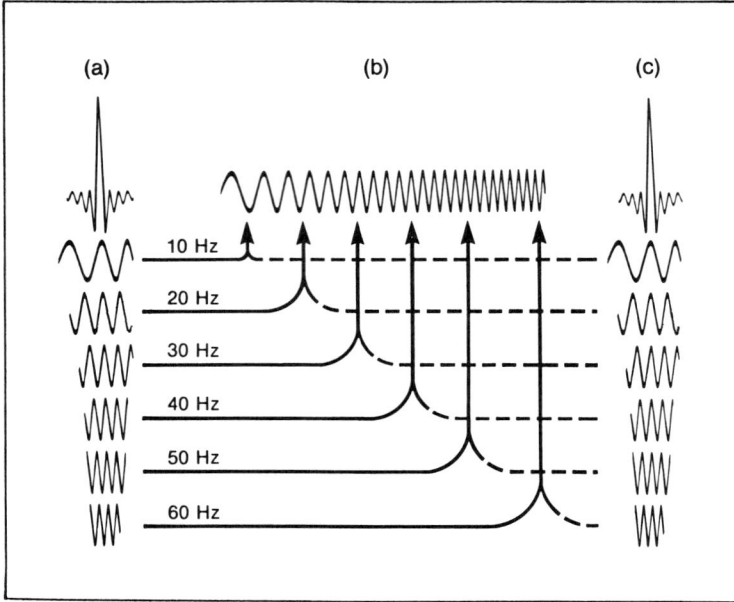

Fig. 2-4 The action of Fig. 2-2 occurs between (a) and (b); that of Fig. 2-3 between (b) and (c).

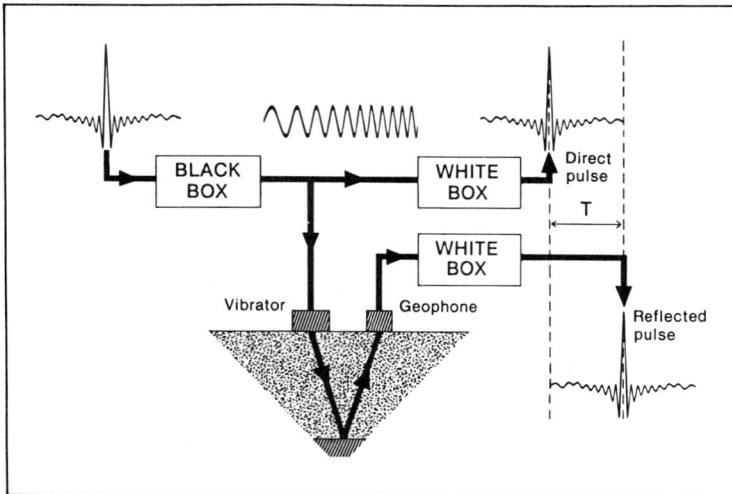

Fig. 2-5 The Vibroseis system in the terms of Figs. 2-2 and 2-3.

The direct pulse (which we shall now call the **zero-time pulse**) is the equivalent of the time break in explosive work. The reflected pulse is the equivalent of a reflection in explosive work, despite the fact that the signal actually transmitted and reflected in the earth was a sweep.

In practice, of course, we have many reflectors and many reflections. If they are separated by a time less than the duration of the sweep, the reflected sweeps overlap when they are detected by the geophone. Given linearity, this overlapping is no problem. It results in the situation of Fig. 2-6. Trace (a) is, we shall say, a trace of a traditional explosive record. It has a time break, two positive reflections (of simplified pulse form), and one negative reflection; the second and third reflections overlap slightly. Now we consider the Vibroseis equivalents. Trace (b) shows the sweep that we generate in the black box and send out into the earth. Trace (c) shows the sweep reflected from the first reflector, delayed by the appropriate travel time. Trace (d) shows the same from the second reflector; like its counterpart reflection on the explosive trace, it is smaller in amplitude. Trace (e) is negative.

The signal detected by the geophone, then, is the sum of traces (c), (d),

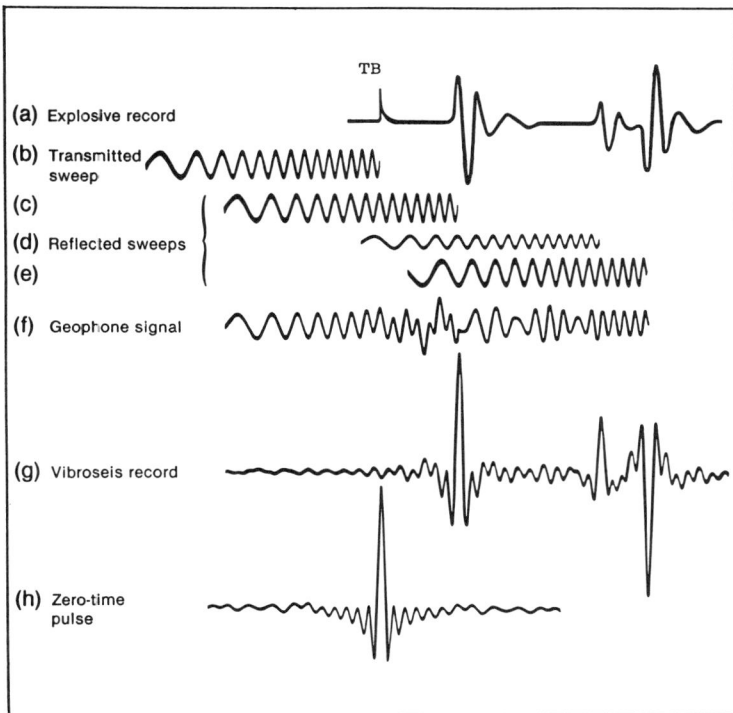

Fig. 2-6 Constituents of a Vibroseis record, corresponding to the explosive record of the first trace. TB (time break) is the time origin of the explosive record.

and (e); this is trace (f). Wherever the reflected sweeps overlap in time, trace (f) is meaningless to the eye—we cannot see specific reflections. But if we take trace (f) and pass it through our white box, out comes trace (g). Each long sweep, despite the overlap, is individually compressed into its own short pulse. If we ignore for the moment our simplification of pulse shapes, the final result is just like the traditional explosive record of trace (a).

Then the zero-time pulse of trace (h), equivalent to the time break of trace (a), is just the result of passing the sweep of trace (b), the sweep that went to the vibrator, through its own white box, as in Fig. 2-5.

The three characteristic ingredients of the Vibroseis system are:

- **A black box to generate the sweep**
- **A vibrator to emit the sweep into the earth**
- **A white box to compress the long sweep into a short reflection pulse**

The result of compressing the sweep sent to the vibrator is the zero-time pulse. The result of compressing the overlapping sweeps detected by the geophone is a train of reflections basically the same as that produced by the traditional explosive method.

Now we should think a little about the ambient noise, which is inevitably being received by the geophone with the sweeps. First, we have said that the peak amplitude of the signal is increased by the white box, as it compresses the sweep into a pulse. Second, we know that the white box is just a system of differential delays (or, in the terms of signal theory, just a phase response) and that as such it cannot change the general level of the noise passing through it. Therefore, the white box has a highly beneficial effect on the ratio of the peak amplitude of the signal to the average amplitude of the noise; it boosts the signal and leaves the noise unchanged.

This fact has given rise to a misconception about Vibroseis and noise. Even today one can still hear people say, as they blithely drive their trucks over the geophones, "Vibroseis is immune to noise." Although we shall see later that Vibroseis rearranges the noise, and has different sensitivity to different types of noise, this unqualified statement is quite wrong.

It is nonsense to say that Vibroseis is immune to noise.

Let us visualize an explosive record and a Vibroseis record having the same ratio of reflection signal to ambient noise. Because of the beneficial effect of the white box on this ratio, it follows that the signal-to-noise ratio *before* the white box (when the signal was in the form of the long low-amplitude sweep) was much poorer. Thus it is normal for a raw Vibroseis trace, as it comes from the geophone, to *appear* to be all noise, with no visible signal at all. Figure 2-7b

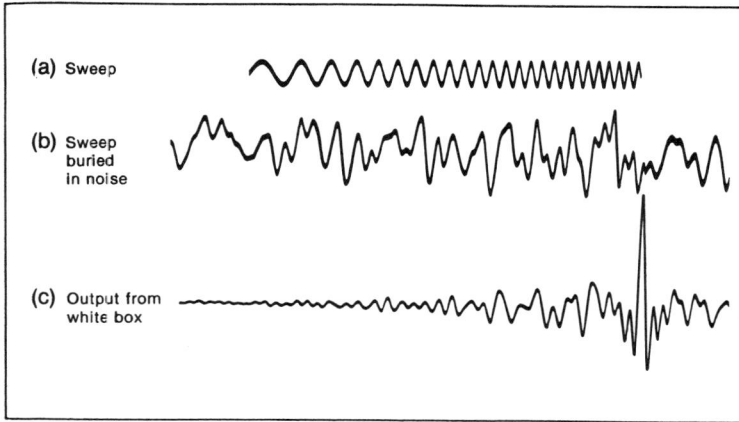

(a) Sweep

(b) Sweep buried in noise

(c) Output from white box

Fig. 2-7   Ability of the Vibroseis process to recover a reflection buried in noise.

illustrates the sweep of Fig. 2-7a buried in noise, and Fig. 2-7c illustrates the improvement wrought by the white box.

We recall from Figs. 2-2 and 2-3 that the black box takes in a zero-phase pulse of defined amplitude spectrum (in our example, 10 to 60 Hz) and puts out a sweep having the same amplitude spectrum, the same energy, much greater duration, and therefore much less peak amplitude.

> **It is the small amplitude of the sweep that allows us to use a surface source without damage. It is the long duration of the sweep that allows us to get the necessary energy into the ground.**

For a defined maximum amplitude of sweep at the vibrator, therefore, our way of increasing the energy is by increasing the duration of the sweep. The white box still outputs the same compressed pulse, but its peak amplitude improves with the duration of the sweep. Figure 2-8 illustrates this.

> **Other things being equal, the longer the sweep the better.**

There is one more conceptual step before we start to formalize these matters. In Figs. 2-2 and 2-3 we stressed, at the bottom of the figures, that the input pulse, the sweep, and the output pulse all have the same amplitude spectrum, flat from 10 to 60 Hz. We obtain no benefit to the signal from components outside this range. The geophone, however, may be receiving ambient noise with frequencies outside this range; these frequencies therefore increase the noise and degrade the signal-to-noise ratio. Clearly, the white box *should include a frequency filter to suppress all frequencies below 10 Hz and above 60 Hz.* Our final view of the white box, then, is that of Fig. 2-9; it includes a rectangular band-pass filter, as well as the differential delays.

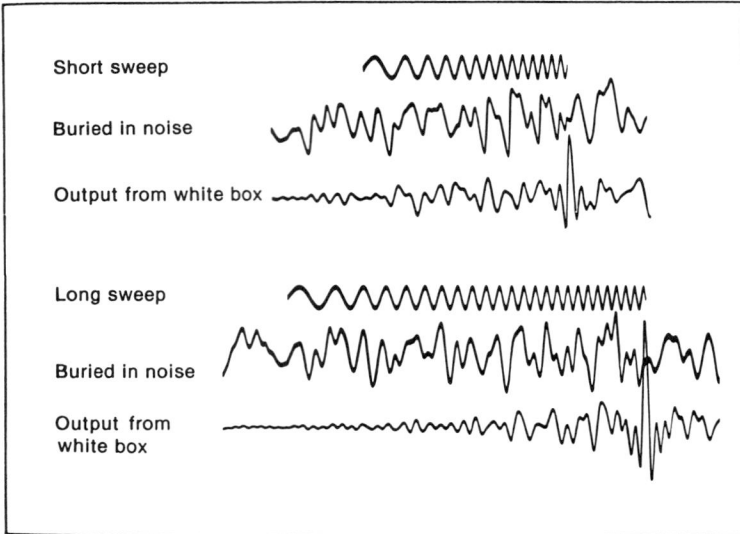

Fig. 2-8 Signal-to-noise enhancement as a function of sweep duration.

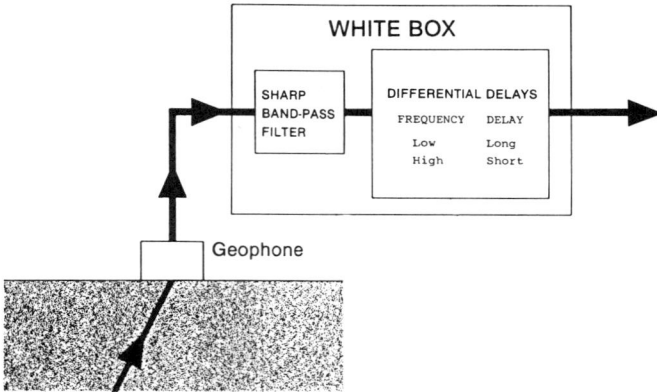

Fig. 2-9 The Vibroseis process needs to provide a band-pass filter, as well as the frequency-dependent delays.

CHAPTER 3

Correlation

Enough of black boxes and white boxes. Now that we are comfortable with the concepts, we should see how it is done in practice.

The entire function of the black box is performed by a digital **sweep generator.** The pure generated sweep[1] is variously called the **control** sweep, the **reference** sweep, the **master** sweep, or the **pilot** sweep. The generator allows choice of the **start frequency,** the **end frequency,** and the sweep **duration** (Fig. 3-1). Typical start frequencies are 4 to 20 Hz, typical end frequencies 40 to 100 Hz, and typical durations 10 to 30 s. The amplitude of the sweep is normally constant; the frequency normally increases linearly with time (an **upsweep**). We shall discuss variants later.

The entire function of the white box is performed in the computer by the process of *cross-correlating the received signal against the transmitted signal.* This is the part that gives trouble to many newcomers to Vibroseis; it is why we have been careful in the foregoing pages to establish the principle of Vibroseis, and to become comfortable with the method, without ever mentioning cross-correlation. Indeed, today we could practice Vibroseis without ever invoking cross-correlation. However, cross-correlation is the traditional way to do it, and we should be at ease with the technique.

The fundamental operation, in correlation techniques, is one of assessing the **degree of similarity between two waveforms.** This measure of similarity is

---

[1]Formal definition in item 1 of Appendix A.

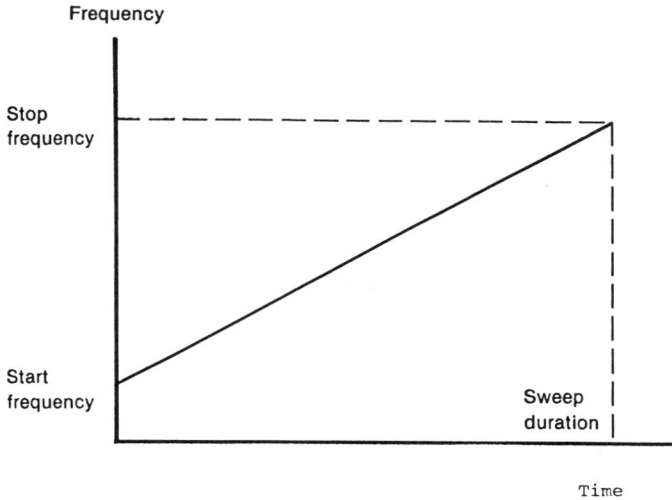

Fig. 3-1 Frequency–time relation of a linear upsweep.

given by multiplying corresponding pairs of ordinates on the two waveforms (a_1b_1, a_2b_2, . . . in Fig. 3-2) and by summing all the products into a single number. If one of the two waveforms is held stationary while the other slides past it, we can compute the variation of the sum of products as a function of the shift between the two waveforms. This is the **cross-correlation function.**[2] As one waveform slides past the other, each waveform is *trying to find itself in the other.* Where it succeeds, the cross-correlation function has large values; where it fails, the function has small values. If the two waveforms are identical, the cross-correlation function becomes the **auto-correlation function.**[3] This function is symmetrical about its maximum value at zero shift.

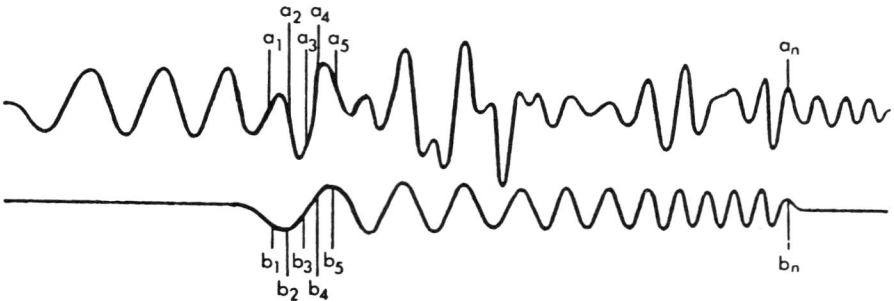

Fig. 3-2 Process of correlation.

[2]Formal definition in item 2 of Appendix A.
[3]Formal definition in item 3 in Appendix A.

Let us now illustrate the correlation process in the light of our previous conclusion: that the entire function of the white box can be achieved by cross-correlating the received signal against the transmitted signal.

We consider first the upper white box in Fig. 2-5, the one whose function is to compress the transmitted sweep into the zero-time pulse. As the sweep is sent to the vibrator, it is also sent to the box; there it is laid down in storage. Then we slide past it a facsimile of itself. We are constructing the auto-correlation function of the sweep. For every position of the slide, we compute the sum of products. In Fig. 3-3, trace (a) represents the sweep laid down in storage and trace (b) the facsimile sweep as it starts to overlap on its way past. For each such position along the way, we compute the product of each pair of ordinates and sum them over the length of the stationary sweep. For some pairs of ordinates (p and q, for example) the product is strongly positive; for others (r and s) the product is strongly negative. The sum of the products ($pq + rs + tu$, plus all the other products) is very small. For this value of the shift between the sweeps, we plot it as such at A on trace (f), which is the auto-correlation function.

A little inspection shows us why this value is small. It is small because the waveforms are so different in frequency; this situation dictates that for every positive product there is always a negative product, and the sum of the products is therefore small or zero. And it leads us to a very useful conclusion:

There is little or no correlation between waveforms of different frequency. The auto-correlation function of a sweep therefore has small values at large shifts.

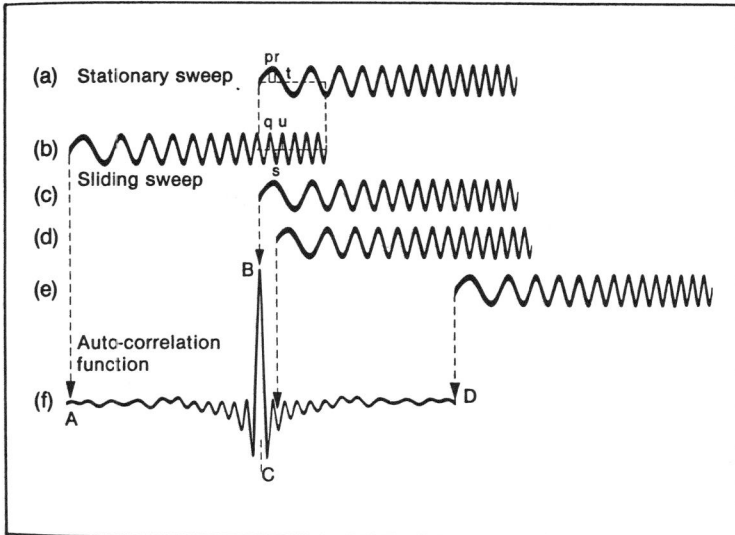

Fig. 3-3 Process of correlation exemplified by the calculation of the auto-correlation function.

In trace (c) we catch the incoming sweep as it is just sliding through the position of coincidence with trace (a) (that is, the position of zero shift between the waveforms). Every pair of ordinates has the same sign, every product is positive, and the sum is therefore strongly positive; this is the zero-shift (or zero-lag) value B of the auto-correlation function.

The zero-lag value of the auto-correlation function represents the energy in the sweep.

Just before and after the coincidence position, there is a systematic tendency for some frequencies to be in antiphase. In trace (d) we catch the sweep where the lowest frequencies are in antiphase and all other frequencies are contributing nothing coherent; the value C is therefore negative.

The position of the sweep on trace (e) [symmetrical with trace (b)] confirms the symmetry of the auto-correlation function; point D is the same as point A.

And now we notice the surprise: trace (f), the auto-correlation function of the sweep, has the same form as the zero-time pulse of Fig. 2-5. And yet, despite the fact that we have described the processes of Figs. 2-5 and 3-3 in quite different terms, there is really no surprise. Let us see why it is so.

In Fig. 3-3, we say that the low frequencies in the stationary sweep (a) are detecting the low frequencies in the moving sweep, (b) to (e); this is because the correlation process—multiplying and summing—produces a significant output only when the frequencies in the two waveforms are similar. Likewise, the middle frequencies are detecting the middle frequencies, and the high frequencies the high frequencies, but at *different positions along the sweep*. Thus the low frequencies are detected early in the sweep and the high frequencies late in the sweep, so the detection process has built into it the frequency–delay relationship of the sweep. In this sense the action of the correlation is exactly that of the white box of earlier figures: a long delay on the low frequencies and a short delay on the high frequencies. And, at the point of coincidence, all the frequencies present contribute *simultaneously* to the output, just as in the Fourier view of Fig. 2-4.

So the correlation process is our white box. It provides the frequency–delay action that we need to compress the sweep.

In Vibroseis, the zero-time pulse is the auto-correlation function of the sweep. This is always a symmetrical pulse, dying away fairly rapidly each side of the center or zero-lag value; the time origin of a Vibroseis record is at this center of symmetry of the zero-time pulse.

Now we turn to the second white box in Fig. 2-5, the one that takes the signal from the geophone. In the absence of noise, this signal is the superposition of many reflected sweeps—one for each reflector—of appropriate amplitude, polarity, and time of arrival. Figure 2-6 was our model, showing the superposition of three reflected sweeps in trace (f). Here we reproduce that trace as trace (a) of Fig. 3-4. This, we visualize, is the signal from the geophone, laid down in storage in the white box. Now we start to slide the sweep past it, just as we did for the zero-time pulse in Fig. 3-3. But this time we are cross-correlating the received signal against the transmitted signal.

As we come to the coincidence position for the first reflection [trace (b)], we have complete duplication over the first several cycles. This is just like auto-correlation. Over the rest of the sweep, however, there is interference between the first reflected sweep and the second. Does this matter?

Not significantly. In Fig. 3-3 there was virtually no correlation between low and high frequencies of one sweep; now we are saying the same between the high frequencies of the first sweep and the low frequencies of the second sweep. The high frequencies of the first reflection are detected without loss by the correlation process, even though visually they are quite obscured by the low frequencies of the second reflection.

Therefore, at a value of shift representing the travel time of the first reflection, the cross-correlation process emits a pulse to show that it has detected the presence of a sweep. Then it does the same for the second (c) and third (d) reflections; in each case, the presence of the other reflected sweeps has

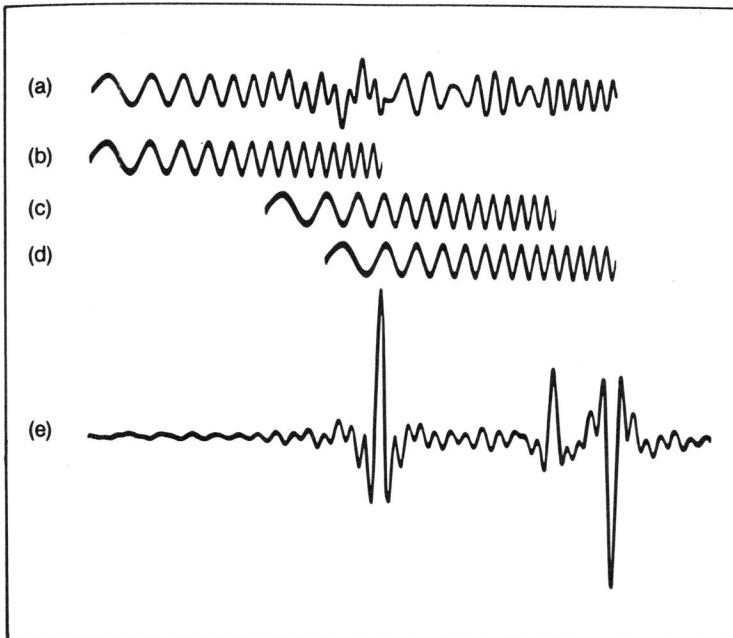

Fig. 3-4 Process of cross-correlating the received signal against the emitted sweep.

no significant effect. The cross-correlation function (e) shows the three reflection pulses.

A Vibroseis record is obtained by cross-correlating the geophone signal against the transmitted sweep. Each resultant reflection—of correct time, amplitude, and polarity—basically has the form of the auto-correlation of the sweep.

As we might guess, the correlation process finds little similarity between the transmitted sweep and the ambient noise. It therefore provides a major improvement of signal-to-noise ratio, relative to the geophone signal before correlation, just as we discussed in Fig. 2-7. But again we must be on our guard against popular myths. We still hear people say that this improvement of signal-to-noise ratio in correlation is an advantage of Vibroseis over the explosive method; this is not true. We remember that the signal emitted in Vibroseis is of low power; the improvement of signal-to-noise ratio is all used up in restoring equivalence to the explosive source.

There is another important feature of the correlation process. We recall from Fig. 2-9 that our white box should have two functions: the compression of the long sweep to a short pulse, and the filtering out of all frequencies not used in the sweep. We have seen how cross-correlation performs the first of these. We have also said that there is little or no correlation between waveforms of different frequency, so cross-correlation performs the second also.

The correlated Vibroseis record contains virtually no frequencies outside the sweep.

This, of course, is very desirable as far as the noise is concerned; now we have only to fight the noise *within the bandwidth of the signal.* But, as we shall see later, it also carries a caution: we must be sure to include in the sweep all the frequencies that we need, for if we do not emit them we cannot recover them.

We have been at some pains, in the foregoing material, to understand the basic concept of Vibroseis and to remove the mysteries of correlation. This will stand us in good stead later. Let us round off this chapter by noting that Vibroseis correlation can be viewed in three ways, which are all strictly equivalent:

1. As shown in Fig. 3-5, the process provides:

   • A steep-sided band-pass filter, passing only the sweep frequencies
   • A set of frequency-dependent delays, by which all the constituent frequencies are brought into alignment and added together in phase

In this view, the output is a zero-phase pulse having a rectangular spectrum with the same limits as the sweep.

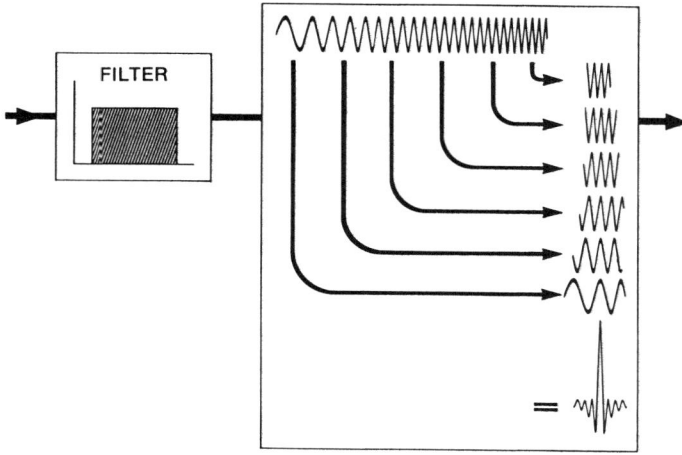

Fig. 3-5   The correlation process provides the desirable band-pass filter and the necessary frequency-dependent delays.

2. As shown in Fig. 3-6, the process *searches* for the sweep by assessing the similarity between the geophone signal and the sweep. In so doing, it is looking for the whole sweep; it is able to find it because the sweep is a *long, highly distinctive* waveform, to which noise is dissimilar. Only when a sweep is found do all the products have a tendency to be systematically positive. In this view, the output is a blip to say, "I have found the sweep, and I mark it thus."

3. The process may also be viewed as a matched filter for the sweep. This concept, discussed in item 5 of Appendix A, is valuable, because it assures us

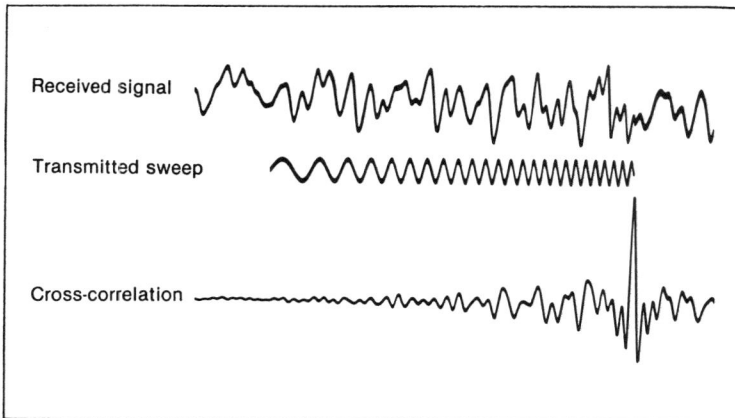

Fig. 3-6   The correlation process may be viewed as a search for the sweep in the received signal.

that there is *nothing better* than correlation to detect the reflected sweeps. The matched filter is the optimum filter when we are searching for reflections in a background of intense white noise.

All three views of the correlation process drive home the point that Vibroseis is not immune to noise. There is a *limited* benefit from correlation: the elimination of noise outside the bandwidth of the sweep, and the compression of all the energy in the sweep into a short pulse. In any real situation, this benefit is numerically defined and numerically limited. More than this we cannot ask.

> **Correlation is the best method of detecting the reflected sweeps; it is remarkable, but it cannot work miracles. With Vibroseis, as with anything else, the less noise the better.**

CHAPTER 4

The

Vibroseis Wavelet

In all the preceding we have ignored all frequency-dependent agencies between the sweep generator and the correlation process; we have assumed that the vibrator, the earth, the geophone, and the instruments all pass the sweep perfectly. In this case, as we have agreed, the reflection pulse has the same form as the zero-time pulse, which is the auto-correlation function of the sweep. (This arrangement would be very acceptable to the interpreter, who would be able to see the shape of the reflections by looking at the zero-time pulse.) We shall stay with this for the present, and explore how the shape of the auto-correlation function depends on the sweep.

We should specify the sweep first. For the moment, we limit ourselves to **linear** sweeps, in which the instantaneous frequency is a linear function of time (Fig. 4-1a). Also for the moment, we assume that the sweep has constant amplitude. Figure 4-1b shows such a sweep. For the sweeps used in practice, the combination of linear change of frequency and constant amplitude guarantees a substantially flat amplitude spectrum between the end frequencies of the sweep (Fig. 4-1c). This is very reasonable, since the linear sweep clearly dwells in the vicinity of each frequency for the same time and at the same amplitude.

Now to the auto-correlation function. Three factors[1] define the auto-correlation function:

- **Amplitude:** This, as was suggested earlier, represents the energy in the sweep. In the following material, where we are concerned with the **shape**

[1]Details in item 4 of Appendix A.

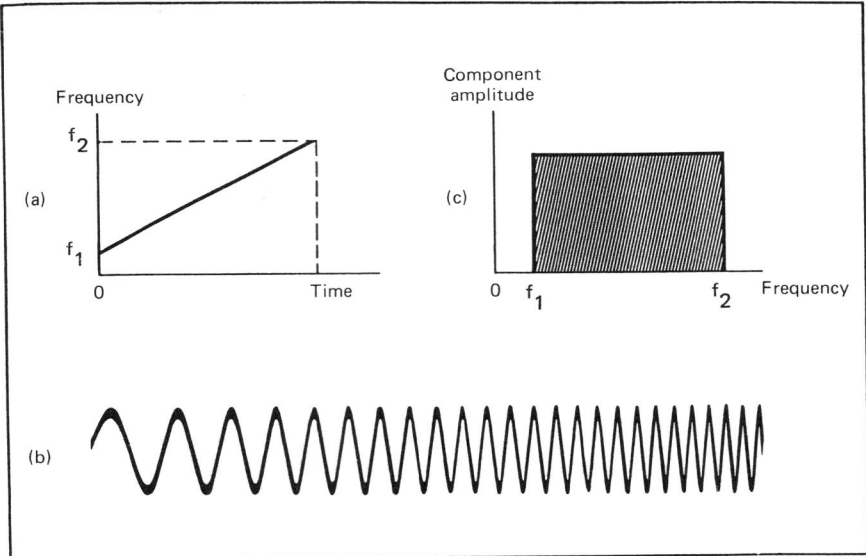

Fig. 4-1   Linear sweep in time and in frequency.

of the auto-correlation function, we shall normalize all the displays to the same amplitude. However, we should never forget that the amplitude of a Vibroseis reflection depends on the sweep energy—that is, on the sweep power times its duration. The more power the better; the longer the better.

- **Envelope:** This has a maximum at zero lag, comes down to zero, bounces up again, back to zero and up again, and eventually dies away (Fig. 4-2a). The time to the first zero, and between subsequent zeros, is the reciprocal of the bandwidth of the sweep.
- **Cosine wave:** This has a peak at zero lag and a frequency that is the center frequency of the sweep (like Fig. 4-2b), but its amplitude stays within the envelope (Fig. 4-2c).

Every linear constant-amplitude sweep has an auto-correlation formed in this way. Looking at the central part, we can always see the cosine wave having the center frequency of the sweep. And we can always sense the effect of the envelope, modulating the amplitude; the greater the bandwidth of the sweep, the more quickly it dies away. This, of course, is what we would expect for a zero-phase pulse; the greater the bandwidth, the shorter the pulse. Let us look at some examples.

Figure 4-3 shows the effect on the auto-correlation function of varying the sweep while *keeping constant the bandwidth in hertz*. Because the bandwidth is always 30 Hz, the envelope is always the same, with zero values at $\pm 1/30$,

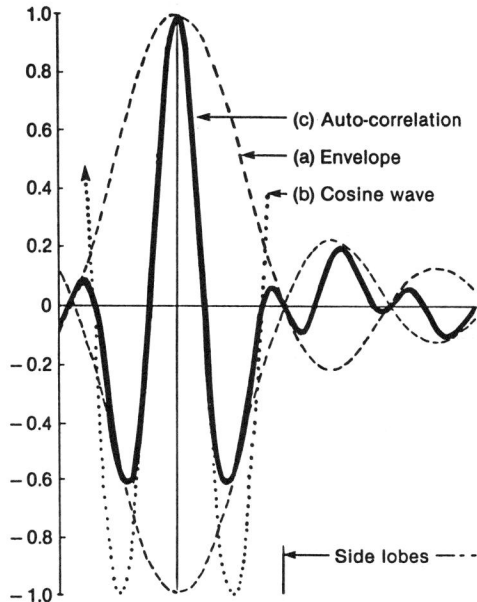

Fig. 4-2 Characteristics of the auto-correlation function.

$\pm^2/_{30}$, $\pm^3/_{30}$, . . . seconds. The center frequency, however, changes; we can see how the frequency of the cosine wave increases from 25 Hz (for the 10- to 40-Hz sweep) to 55 Hz (for the 40- to 70-Hz sweep).

Figure 4-4 shows us the effect of varying the sweep while *keeping constant the center frequency.* The cosine wave is always at 40 Hz; the envelope exhibits a faster decay as the bandwidth increases from 27 Hz (for the 27- to 54-Hz sweep) to 63 Hz (for the 9- to 72-Hz sweep).

Finally, Fig. 4-5 shows us the effect of the bandwidth in octaves. Auto-correlations of sweeps having the same bandwidth in octaves all have the *same shape;* they are just spread out differently in time. A greater bandwidth in octaves yields a spikier pulse, with less overshoot.

It is worth looking at all these auto-correlations again. First, their message is important to us when we go to the field to solve a particular geological problem. Second, there is another important conclusion lurking in them. We see that, although the center frequency is always seen quite clearly near the high-amplitude part of the auto-correlation function, the effect of the modulating envelope makes two other frequencies appear—the end frequencies of the sweep. This is particularly clear, for example, on the 9- to 72-Hz sweep of Fig. 4-4.

The effect occurs because, in Fourier synthesis, destructive interference at one frequency can be effective *only where the components on both sides of that frequency are present.* If the spectrum terminates abruptly, the pulse always contains a superabundance of the end frequency.

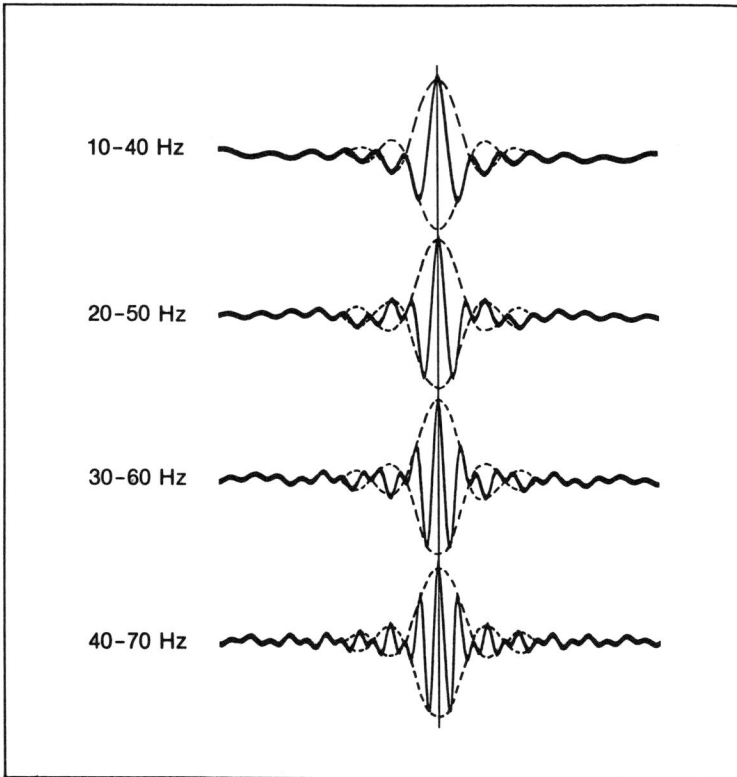

Fig. 4-3 Auto-correlations of sweeps having a constant bandwidth in hertz.

This gives us the key to improving the auto-correlation functions by making them less ringy: we must round the shoulders of the spectrum.

The side lobes of the auto-correlation function of a sweep can be reduced by tapering the ends of the sweep.

Figure 4-6 compares the auto-correlation functions of a sweep with and without a modest tapering; the sweep amplitude is tapered to zero over a few hundred milliseconds at each end of the sweep.

We pay a price for this improvement, of course; the bandwidth is decreased. But, on balance, the effect is clearly desirable.

So far so good. Now we are ready to consider the real world—specifically, the frequency-dependent effects of absorption and other agencies in the earth (the **earth filter**). Up to this stage, we remember, we have accepted that the only frequencies that do us any good lie within a defined band (10 to 60 Hz in the

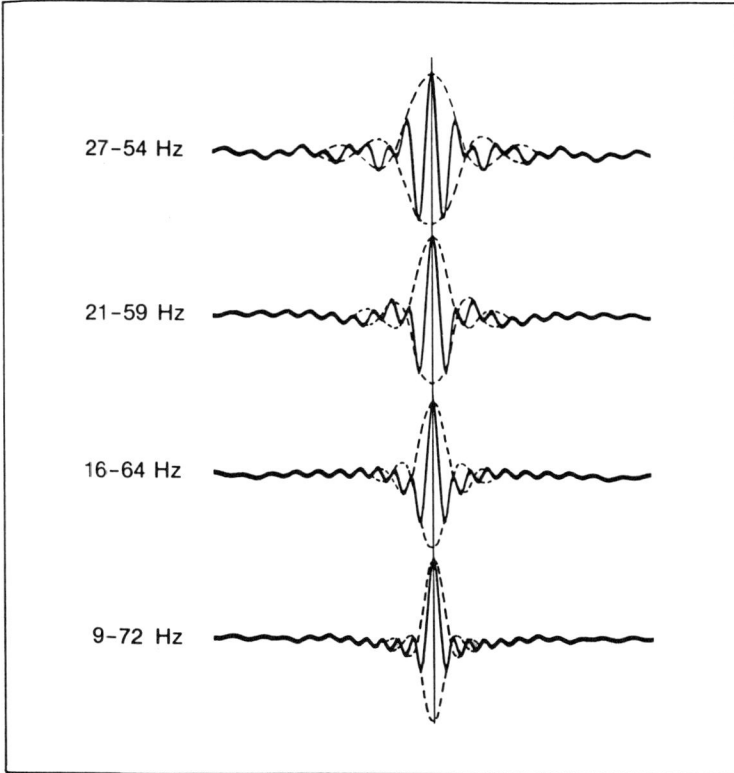

Fig. 4-4   Auto-correlations of sweeps having a constant center frequency.

example of Fig. 2-1), but we have assumed that the earth has a flat passband within this range. Now we must acknowledge that, in fact, the response of the earth filter is likely to be quite sharply peaked within this range.

Let us suppose that this response is as shown in Fig. 4-7a. As we have said before, it yields no significant output below 10 or above 60 Hz; now we are adding that it is strongly peaked at about 30 Hz. In the explosive case the input to this system approximates the spike of Fig. 4-7b; however, as we have agreed, it would not make any difference if the input were the pulse of Fig. 4-7c, the zero-phase pulse of rectangular bandwidth from 10 to 60 Hz. The output is the reflection pulse of Fig. 4-7d: the pulse with the peaked amplitude spectrum of Fig. 4-7a and (we shall assume) minimum phase.

In the Vibroseis case the transmitted sweep (Fig. 4-8a) passes through the earth response in the same way. The reflected sweep therefore has the form of Fig. 4-8b. The amplitude response of the earth can be seen directly in the amplitude variation of the sweep. The phase response can be seen by compar-

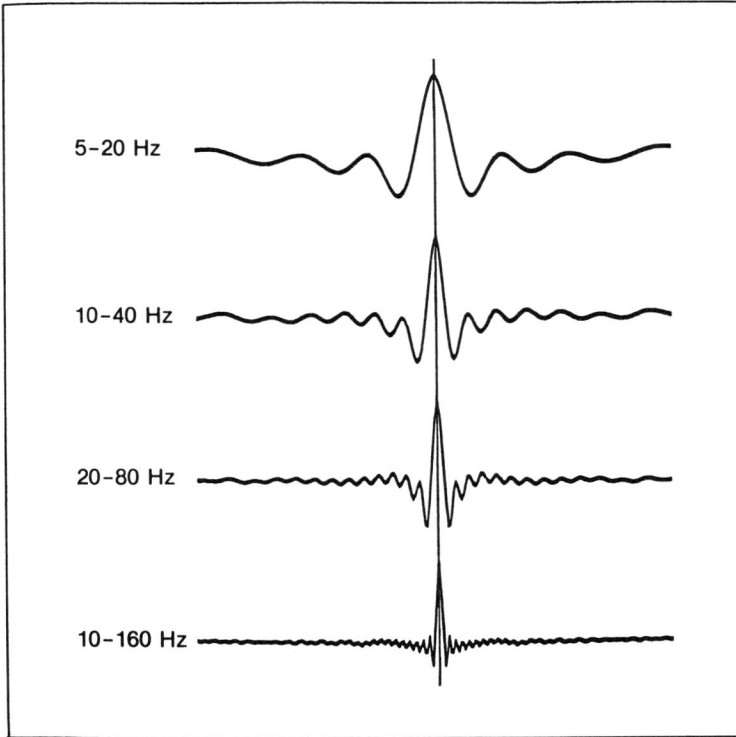

Fig. 4-5 Auto-correlations of sweeps having a constant bandwidth in octaves and (bottom line) an increased bandwidth in octaves.

ing individual cycles; the low frequencies lead, the middle frequencies are in phase, and the high frequencies lag.

Amplitude-frequency responses and phase-frequency responses acquire a delightfully direct significance when they act on a sweep.

We now correlate the received signal (b) against the transmitted signal (a). The result is the cross-correlation function (c), which is identical to the reflection pulse obtained with the explosive method. Figure 4-8c *is* the reflection pulse.

This identity allows us a useful mental view of the Vibroseis process:

The effective input to the earth, with Vibroseis, is the auto-correlation of the sweep.

Or we may say (and this is the same thing):

The reflection pulse is the convolution of the earth response with the auto-correlation of the sweep.

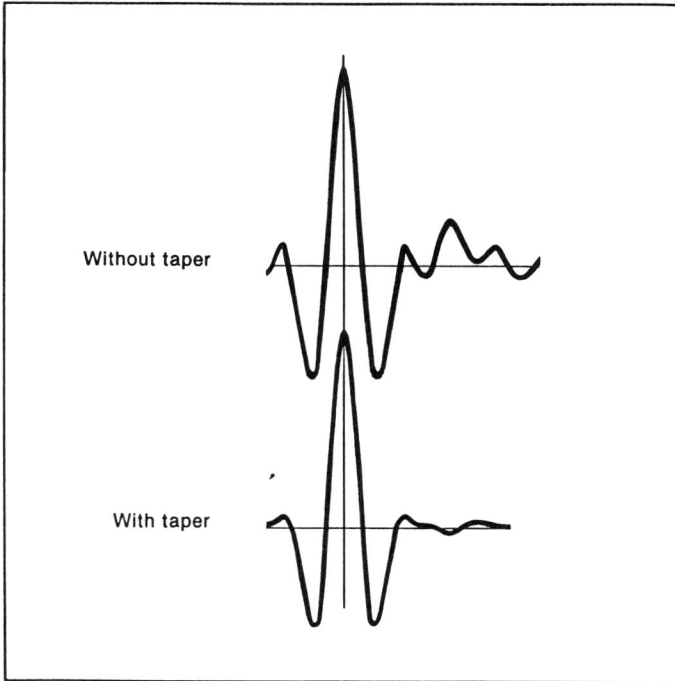

Fig. 4-6 Effect on the auto-correlation of tapering the ends of the sweep.

All we need to do, then, to guarantee the basic equivalence of the Vibroseis and explosive methods is to use a sweep that includes all frequencies passed at significant amplitude by the earth.

Provided the sweep is wide enough, records with Vibroseis and with explosions are conceptually identical.

Of course, they may not be identical in practical detail. The frequency-selective action of the cavity formed by the shot may not be the same as that of the vibrator and its coupling to the ground. A buried shot has a surface ghost. And, for practical reasons, we may choose to use different arrays and different spread geometries with the two methods. Nevertheless, the basic equivalence is a very valuable concept.

The proviso—that the sweep band must be wide enough—is obviously important. This faces us with an operational dilemma. To be safe, we would choose a sweep of very large bandwidth. But we cannot increase the bandwidth while keeping the sweep duration the same, for that would decrease the time spent in the neighborhood of the useful middle frequencies, and therefore the

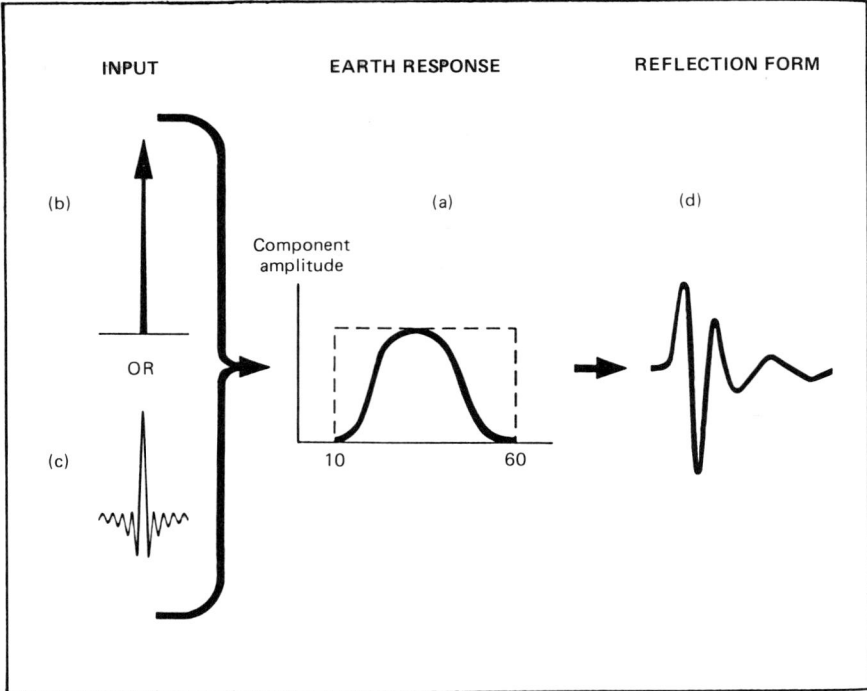

INPUT   **EARTH RESPONSE**   **REFLECTION FORM**

Fig. 4-7 Effect of the earth filter on both spike and Vibroseis inputs.

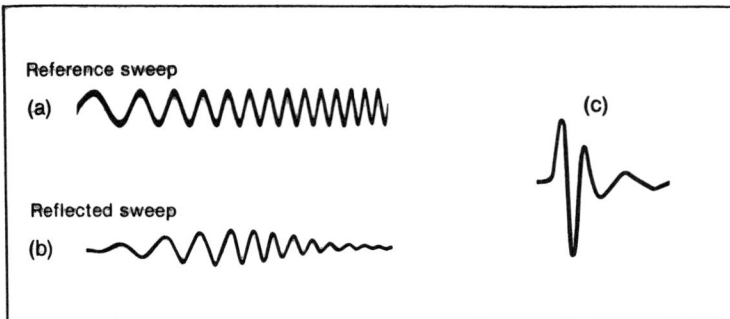

Fig. 4-8 Effect of the earth filter on the sweep itself and on the cross-correlation (c).

useful energy. So we must use longer sweeps. But then we would be spending much of our field time sweeping very low and very high frequencies that do us little good, and that would be very expensive.

The choice of sweep frequency is a compromise between cost and our desire to maintain the greatest bandwidth reasonably allowed by the earth filter.

Let us discuss for a moment the consequence of this compromise for the shape of the Vibroseis reflection. First, there are the consequences for the amplitude spectrum. Figure 4-9 shows a representative situation; we choose a sweep encompassing most of the response of the earth filter at our range of target depths, and we emerge with the truncated spectrum of Fig. 4-9c. Now, what do we need to do about tapering the ends of the sweep for the reduction of side lobes? Nothing. The earth filter has already done more than enough tapering.

Yet most operators still choose to apply some tapering. Partly, perhaps, this is because there are always shallower reflections, for which the earth filter is less peaked (and for which the shoulders of the amplitude spectrum are consequently sharper). And partly, perhaps, it is to cover the case when, for eco-

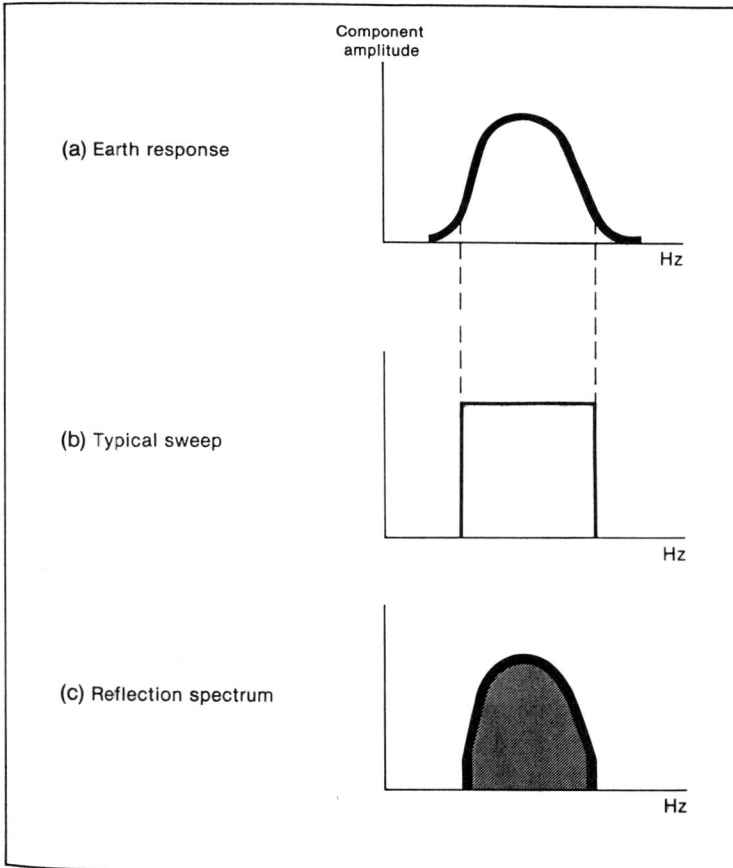

Fig. 4-9 The earth filter ordinarily provides more than enough tapering of the sweep.

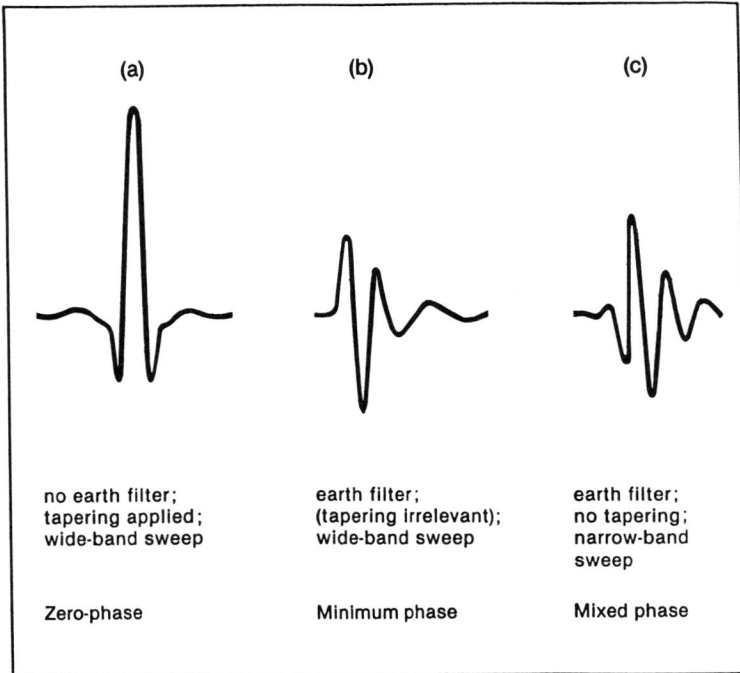

Fig. 4-10   The action of a minimum-phase earth filter on a zero-phase auto-correlation yields a mixed-phase Vibroseis reflection.

nomic reasons, the sweep is very narrow, straddling the peak of the response (so that the shoulders are again sharp).

Since the early days of Vibroseis, it has been standard practice to taper the sweep *sent to the vibrator.* It is true that such tapering is kind to the vibrator; it allows all the motions to build up (and decay) gradually, and so avoids any transient lurch that may be generated if the mechanisms do not respond perfectly to the onset and end of the sweep signal. But the avoidance of starting and ending transients is an objective quite different from the shaping of the spectrum. The conclusion is that we apply a smooth and *very brief* taper at the ends of the transmitted sweep—just to avoid transients, and to be kind to the vibrator—but we do *not* apply deliberate spectrum-shaping tapering to the transmitted sweep. To do so would give a pointless advantage to the noise, at the tapered frequencies.

Except to the degree necessary to avoid mechanical transients, we should do any desired sweep tapering in the processing—not at the vibrator.[2]

[2]See also item 7 of Appendix A.

Second, there are the consequences for the phase spectrum. We have said that if there were no earth filter the Vibroseis reflection wavelet would be zero phase; any side lobes remaining after tapering would be symmetrical (Fig. 4-10a). But in practice there *is* an earth filter, and we have assumed its response to be minimum phase. We have also said that, if the sweep is wide enough, the phase spectrum of the earth-filtered wavelet is zero for the Vibroseis process plus minimum phase for the earth, which is minimum phase. In this situation the wavelet builds up fairly quickly to its maximum amplitude, with most of the side lobes *following* this maximum (Fig. 4-10b). So what happens when we compromise, and adopt a smaller bandwidth for the sweep?

For practical compromises of sweep bandwidth, the Vibroseis wavelet is mixed phase.

The effect of this is that the wavelet is asymmetrical, but not as much so as a minimum-phase wavelet. In particular, some weak side lobes may persist *before* the pulse maximum (Fig. 4-10c). These are not ordinarily evident on reflections, but they are often seen (particularly if aggravated by an automatic gain control) on the first breaks of reflection records and borehole velocity surveys.

In later material we shall continue to speak of the Vibroseis reflection as basically a symmetrical auto-correlation function, modified by the minimum-phase earth filter. This is a convenient way to visualize it. As we do so, however, we shall bear in mind that we may choose to manipulate this wavelet to other shapes in practice (item 12 of Appendix A).

After all that, we need a break. Let us go to the field.

CHAPTER 5

Vibroseis
in Action

Figure 5-1 illustrates a vibrator. Toward the rear sits an engine, roaring away at the threshold of pain; it furnishes the prime power. In the center is the baseplate (or **pad**), lowered into contact with the ground and ready to vibrate. In the cab sits the vibrator operator, bored stiff; today he sends out his millionth sweep. Lurking somewhere out of sight is the vibrator technician, whose job it is to repair this great mass of machinery whenever it succeeds in shaking itself to pieces.

Just under the right arm of the vibrator operator is the sweep generator (Fig. 5-2). This is our black box of Fig. 2-2; to remove the air of mystery the manufacturer has painted it cream. Under the flip-up armrest are the controls for start frequency, stop frequency, and sweep length; these can also be selected remotely from the recording truck, thus avoiding the disaster (it has happened) of different vibrators sending out different sweeps.

Different vibrators? Yes, because it usually makes economic sense to employ three or four vibrators (Fig. 5-3), rather than one. The initial cost is enormous, but the increased power reduces the time necessary to inject sufficient energy into the earth, and so improves the production. Further, operations can continue when a vibrator goes sick. (With vibrosis.)

The geophone and spread layout is basically conventional. However, there may be some changes to minimize the effect of the stronger surface waves generated by a surface source. Typically, these involve a longer offset from the source to the near geophone array and (for traditional array design) longer

Fig. 5-1   Representative vibrator (courtesy of Mertz Inc.).

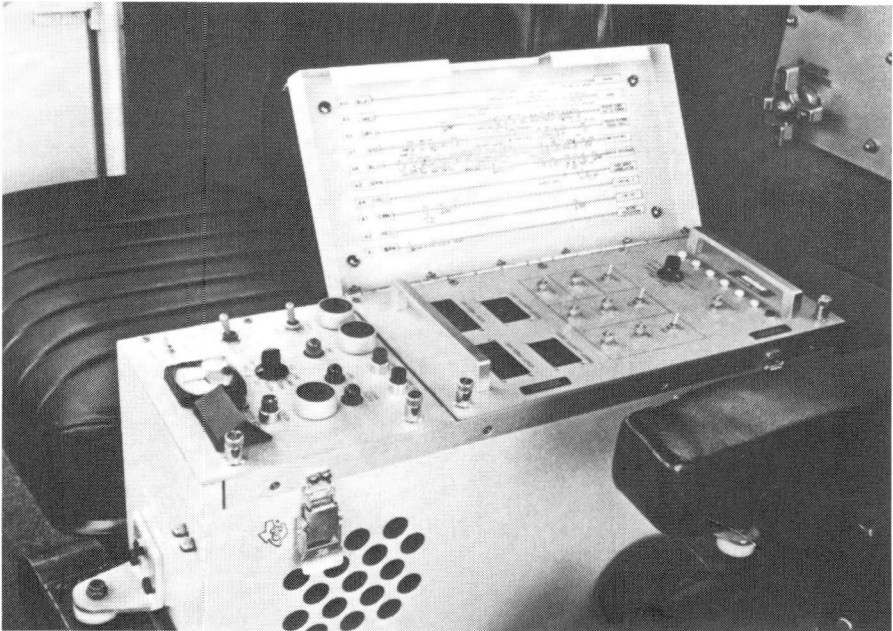

Fig. 5-2   Sweep generator in the vibrator cab (courtesy of Texas Instruments).

Fig. 5-3   Three vibrators at work. The baseplates are down and sweeping.

geophone arrays. Figure 5-4a shows a convenient practical arrangement: 48 channels, an array length equal to the group interval, a straddle spread, and a 2½-interval offset from source to near array. This arrangement can roll along, of course, in the manner described in Chapter 1.

Then Fig. 5-4b enlarges the vicinity of the source point SP (also called vibration point, or VP). The triangular flags represent the array centers, as usual. The effective source point is halfway between flags.

One technique is to start with one vibrator at the flag on the back side, and the others (the figure suggests a total of three) nose to tail in front of it. All is ready for the first sweep; everything is tested, and the pads are down. The command signal comes by radio from the recording truck. Brrrrrr - rrrrr - rrrr - rrr - rr - r. Standing by the side of the vibrator, we hear the engine change note as the system draws power. At the low frequencies we can actually see the baseplate (and the ground) moving; at the high frequencies we sense it mostly through our feet. Perhaps we are working in a fairly difficult area, requiring much energy; the sweep lasts 24 s.

Within a split second of the end of the sweep, the vibrator operator raises the pad and drives forward. (When the vibrators are working nose to tail, as here, the operator with the fastest reactions takes the lead vibrator, for obvious reasons.) In this example, the three vibrators drive forward one-twelfth of the group interval. Then the pads go down, and within 6 s everything is ready for

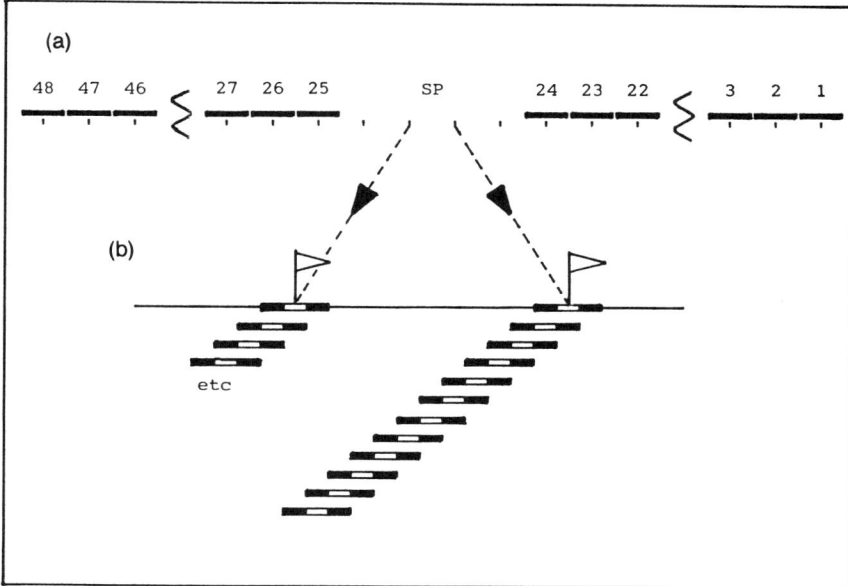

Fig. 5-4 The emission of multiple sweeps from each of three vibrators (black–white–black) across a source array. The diagram shows the source array extending from flag to flag; since the geophone groups are centered on the flags, this means that the effective source point (the center of the source array) is halfway between the groups. This arrangement has significant advantages, discussed in Chapter 10. The vertical dimension in the diagram merely allows the successive in-line positions of the vibrators to be shown without overlap.

the next sweep. Brrrrrr - rrrrr - rrrr - rrr - rr - r. And so on . . . 24-s sweeps and 6-s move periods, until 12 sweeps have been emitted. This brings the three vibrators to the start position for the next SP.

Meanwhile, of course, the observer has been vertically stacking the geophone signals from the 12 sweeps. This is done by switching the signals into 30-s segments and stacking the 12 segments. In effect, then, the observer is forming (for each geophone array) a single 30-s signal that represents a **source array** equal in length to the group interval. This may be helpful for further attenuation of the surface waves. The single signal may then be correlated, or recorded on tape for subsequent correlation.

Since everything is now in position for the next SP, the sequence continues exactly as above. This regularity is part of the appeal of Vibroseis; if there are no malfunctions, the progression can be almost as smooth as a marine operation. In our example, each SP takes exactly 6 min, one after the other, after the other, after the other. Which is why the vibrator operators are bored. But it is better than sloshing around a drill truck, or loading dynamite.

Let us look again at the sweep cycle. Figure 5-5a suggests the first three

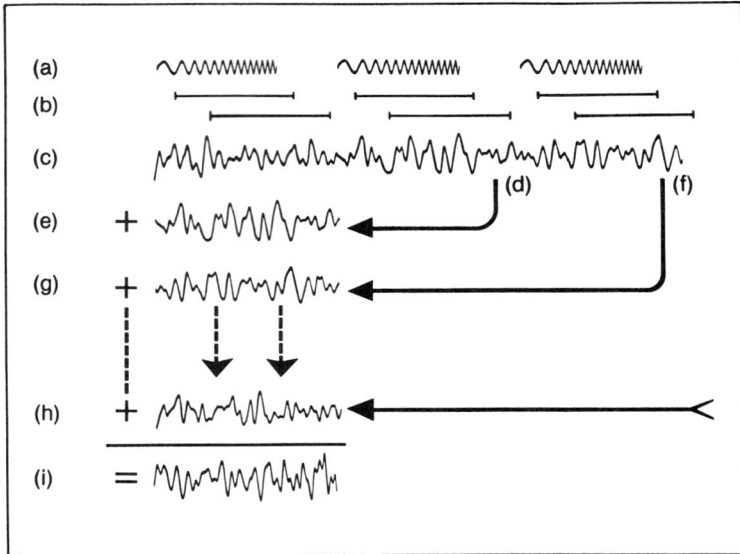

Fig. 5-5 Segmentation and vertical stacking of the Vibroseis signal.

sweeps. In our example, these were each 24 s in duration, separated by 6 s in which the vibrators moved to form the source array. In Fig. 5-5b we see representative time positions for an early reflection of the sweep and for a late reflection (at almost 6 s). The actual signal received by the geophone is the sum of all such reflected sweeps, plus the ambient noise; this might appear as in Fig. 5-5c.

We could, if we wished, record this signal directly on tape. Since the Vibroseis operation (as exemplified above) is continuous, this signal is continuous; the tape would just run from beginning to end, and another tape would start on a second transport. This would save a stacker, but use a great deal of tape (and a great deal of input time in the processing computer). So normally we use a stacker. The first 30 s of the continuous geophone signal enters the stacker; then the second 30 s (Fig. 5-5d) is aligned with the first 30 s (e) and added to it. The third 30 s is similarly aligned, (f) and (g), and added to the sum. The process continues to the twelfth 30 s (h), at which time the final vertically stacked signal is available (i) to be recorded on tape for subsequent correlation. This signal is 30 s in length; in our example, one such signal (representing one SP) is recorded every 6 min.

During the subsequent correlation, of course, this 30-s signal is correlated against a facsimile of the 24-s transmitted sweep. The correlation is performed from 0 to 6 s of shift between the waveforms, thus yielding a 6-s reflection record.

It has become common to refer to the period between sweeps as the "listening period." This is quite wrong.

The principal function of the period between sweeps is to allow the vibrators to move (and so to form a source array). This period has no special significance as a listening period; the whole cycle is the listening period.

It so happens that 6 s is a practical time for moving the vibrators far enough to form a source array. It also happens that 6 s is the usual length for seismic records used in petroleum exploration. This is coincidental, and must not confuse our thinking. For example, we may be working over shallow basement, where 3-s records are sufficient. We still use 6 s between sweeps, to allow time to move the vibrators. In the correlation, however, we vary the shift only over 3 s, instead of 6 s.

What do we do if we cannot move the vibrators far enough in 6 s? One option is to allow as long as it takes, 24 s if necessary. This is not a good solution. We have accepted, in our example, that a usable signal-to-noise ratio requires the energy of twelve 24-s sweeps; if we use a 6-s move period, this takes 6 min per SP, but if we use a 24-s move period, it takes nearly 10 min. The cost of the survey is significantly increased.

A better solution is illustrated in Fig. 5-6. Instead of placing the three vibrators nose to tail, we separate them and give each one-third of the array. Now they have only a small distance to move between sweeps, but a longer distance between source points.

For emphasis, we can consider the case where the move period is reduced to less than 6 s, but where we still need 6-s records. In principle, this is perfectly feasible; but, in practice, we would probably say that the small improvement in production would not warrant all the hard thinking we would have to do.

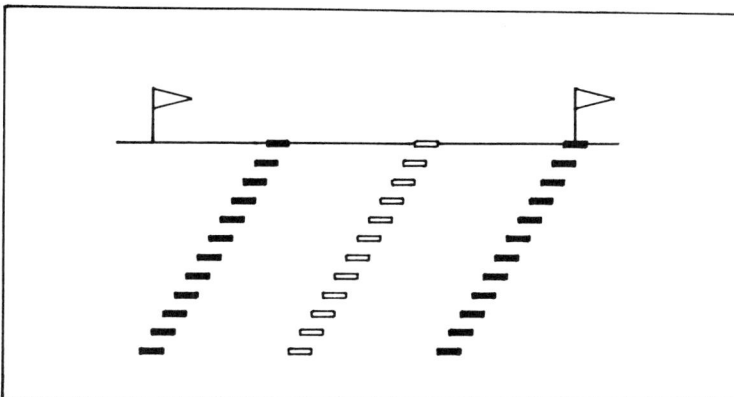

Fig. 5-6 Alternative method for building up a source array.

Given that signal-to-noise considerations require a certain sweeping energy (and hence a certain sweeping *time*) per SP, why do we not use just one
very long sweep per SP? The elimination of all but one of those move periods
would improve the production (to about 5 min per SP) and would no longer
need a stacker. There is nothing wrong with the idea; it makes no difference to
the signal-to-noise ratio if we vertically stack 12 sweeps of 24 s each or use only
one of 288 s. The stacking of 12 sweeps increases the signal by 12, the noise
(which adds incoherently) by $\sqrt{12}$, and the signal-to-noise ratio, therefore, by
$\sqrt{12}$. The use of a single sweep 12 times longer increases the sum of products for
the signal by 12, and for the noise by $\sqrt{12}$. The two situations are identical. So
why not? Because the single long sweep strains the capacity of a correlator;
furthermore, the response of the source array given by occupying many positions is sometimes more desirable than that given by just three or four vibrators
stationary throughout one long sweep.

The point is still important, however. In the early days of Vibroseis,
hardware considerations dictated a 7-s sweep and a 6-s move period; this was
inefficient. Today, in any area of problematical signal-to-noise ratio, we tend to
use the longest sweep that the equipment can handle. Only if this means too
long a move for the vibrators (and hence a poor source-array response) do we
reduce the length of the sweep. In practice, these factors mean that we usually
employ sweeps of 18 to 26 s for difficult areas and 12 to 18 s for easier areas.

Ordinarily, as we have said, the sweep cannot be seen in the raw signal
from the geophone; all we see is noise. Even on the vertically stacked signal we
may see only the dimmest suggestion of sweeps. If we do see the sweep
character clearly, above the noise, it probably means that the surface waves are
serious; what we are seeing is the sweep that has traveled to the geophone as a
surface wave. Therefore, we *prefer* to see no sweeps. This is part of the problem
with the old myth about noise in Vibroseis. An observer can always arrange
that the uncorrelated record shows no sweep merely by allowing more noise on
the line (by jumping up and down on the geophones, if necessary). But that, of
course, is very, very wrong.

Bringing together everything we have learned about Vibroseis, we can see
that it combines an intriguing theoretical idea with a very smooth operation in
practice—and no real terrors of comprehension. It takes advantage of all the
stratagems to implement a surface source: it splits the total required energy into
many small emissions, it emits only those frequencies that are useful, and it
packages the required energy into signals of low power and long duration. So
effective are these devices that the method can work on roads without damage.
This is a large part of the commercial success of Vibroseis, for it eliminates
untold problems of access and crop damage. For many surveys these considerations are paramount, and there is no question that the choice should be
Vibroseis.

However, we seek a balanced view, and in some situations we need to
remember that there are also disadvantages to surface sources:

- As we have already noted, a surface source generates more severe ground roll.

- A dynamite shot below the weathered layer allows us to make a better correction for the near-surface—a better solution to the "statics" problem.

- As surface conditions change, the signal emitted by a surface source changes; with drilled holes, we can usually drill to the same formation, and so maintain a more constant source pulse.

- At present, unsolved problems exist in adapting vibrators to marshes, peat bogs, and other soft conditions.

- Working along roads is a great convenience—true. But sometimes the roads are not in the right place for our targets, and sometimes crooked roads introduce significant doubt into our interpretations. So occasionally we *need* to go cross-country, and the vibrators may not be able to do it.

Furthermore, vibrators are expensive, complicated, and imperfect. Let us look at them.

CHAPTER 6

The Vibrator

The problem, of course, is immense: how to shake the earth at 60, 80, 100 times a second? Even at 10!

We must apply an oscillating force, a very large force. But what is this force to push against? All we can do is to take the heaviest weight we can consider for truck-mounted operations, and push against that.

Let us start with Fig. 6-1, which shows the following components:

- A rectangular baseplate (typically 2 m, or 7 ft, in the direction illustrated, transverse to the vehicle; 1 m, or 3 ft, in the other direction). It is important that the baseplate be rigid; otherwise, there is a danger, particularly at high frequencies, of flexing in the baseplate—so that some of the vibration is not delivered into the ground. It is also desirable that the baseplate be light relative to the mass of the earth that is going to move with it. Baseplate design, therefore, has something in common with aircraft design.

- A means by which the baseplate can be raised and lowered. This may be some sort of frame (as illustrated) driven up and down by hydraulic rams. We remember that we wish this action to be fast, to minimize the move period. Since the cylinders of the hydraulic rams are anchored to the chassis of the truck, actuation of the rams can raise the truck on the baseplate, and so hold the baseplate down with the full weight of the truck. This is desirable to prevent the baseplate from jumping off the

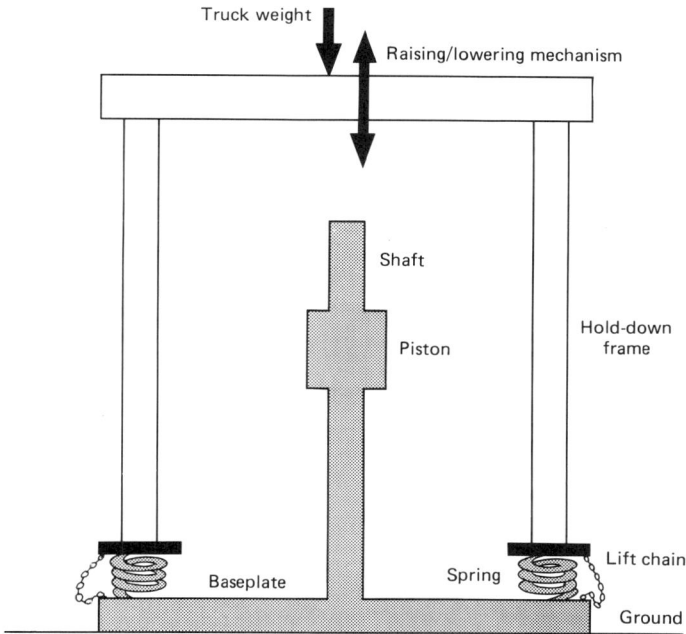

Fig. 6-1 Vibrator baseplate and the arrangement to hold it down.

ground as it vibrates; it is this feature that gives the strange mosquitolike action to a vibrator, and which so fascinates the bystanders. It is also the reason why powerful vibrators require heavy trucks.

• Springs, between the baseplate and the hold-down frame. These are necessary so that the vibrator does not directly shake the truck. Usually they are air bags, of a natural frequency well below the lowest frequency of a sweep. They are bypassed by a slack chain (or equivalent device) so that when the baseplate is raised the air bags are not torn apart.

• A vertical drive shaft anchored very securely to the center of the baseplate, and machined near its upper end to form a piston.

For simplicity of illustration, we can now omit the raising and lowering gear and concentrate on the vibrator itself. We turn to Fig. 6-2. Here we see the heavy weight we push against—the **reaction mass;** typically, this is of solid metal, weighing 2 to 3 tonnes. This reaction mass is free to slide on the shaft; into it is formed a cylinder in which the baseplate piston can slide with tight clearance.

Now our plan is clear. We first pump hydraulic oil into the cylinder above the piston; this simultaneously raises the reaction mass and pushes the baseplate down. Then we withdraw the oil from above the piston and pump oil

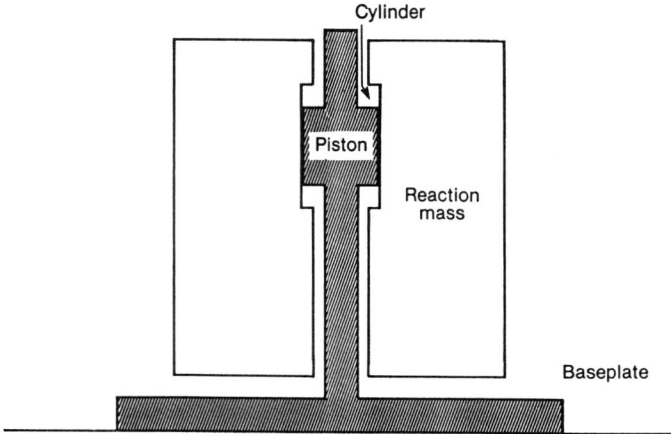

Fig. 6-2   Baseplate, piston, and reaction mass.

below it; this lowers the reaction mass and allows the earth under the baseplate to relax. By alternately pumping and evacuating, we can induce continuous vibration between the reaction mass and the baseplate. We have a vibrator.

The motion is shared, of course, between the reaction mass and the earth. The usual situation is that at low frequencies the reaction mass is not as heavy as we would wish, and there is considerable movement of both the reaction mass and the earth. At high frequencies, however, the reaction mass is almost still, and most of the motion occurs in the earth.

There remains the formidable engineering problem of supplying and controlling the hydraulic oil. Figure 6-3 illustrates the basic scheme. Oilways are drilled through the shaft to connect to the cylinder above and below the piston. These must be wide enough to take the enormous flow of oil, and short enough that the elasticity of the oil does not absorb all the motion at high frequencies; this is why the piston is near the top of the shaft. The two oilways connect to the two outputs of a servo-valve bolted to the top of the shaft. This servo-valve is, in effect, a double two-way switch. It connects each of the outputs alternately to the high-pressure side and the low-pressure side of the oil pump. The action is as though the black bar in the figure were being continuously oscillated back and forth between left and right positions. In the left position the high-pressure side of the pump is connected to the upper cylinder, and the earth is pushed down. In the right position the earth is allowed to spring back.

This servo-valve is really a remarkable device. It can handle an oil flow of as much as 16 liters (l)/s (200 gallons per minute, gpm) and frequencies up to 100 or even 200 Hz. Most models have a main stage (as suggested diagrammatically in the figure), a pilot stage to drive the main stage, and an electric

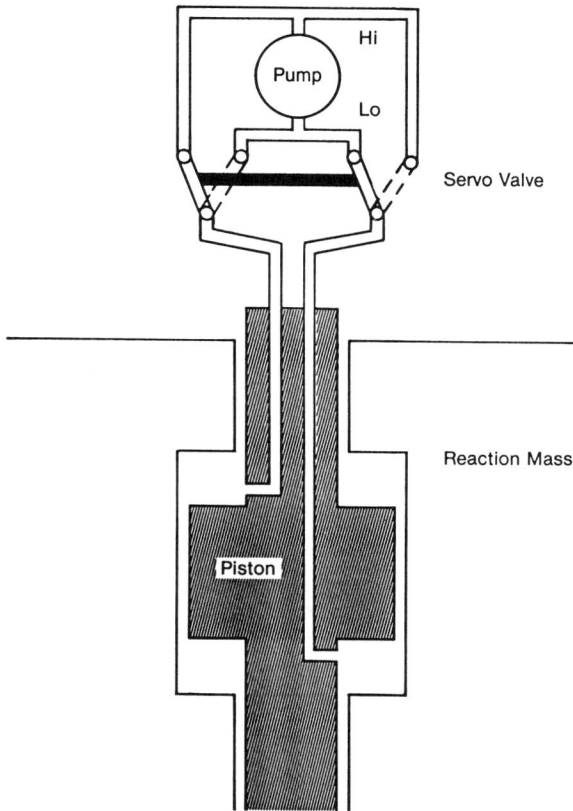

Fig. 6-3 Means for driving the piston.

torque motor to drive the pilot stage. The drive for the torque motor, of course, comes from the sweep generator.

The remaining components in the hydraulic system are the pump itself, an oil tank, several critically important oil filters, an oil cooler (the vibrator is only about 7% efficient), and an auxiliary pump to charge the system. There are also two accumulators or reservoirs, one on the high-pressure side of the pump and one on the low. These supply temporary oil flow beyond the capacity of the pump.

Figure 6-4 shows the baseplate and part of the lift mechanism; Fig. 6-5 is a general view of the vibrator housing.

The mechanical and hydraulic components of a vibrator work very hard. That enormous reaction mass must be constrained, as the vibrator truck rocks and sways and bumps over rough ground. The hoses must flex as the baseplate

Fig. 6-4   One end of the baseplate, showing also a hold-down ram, the air bags, and the limiting mechanism equivalent to the lift chain.

goes up, down, up, down, up, down. And the vibrator and the baseplate are subjected to something like 1 million stress cycles every day. We should spare a moment to salute the engineers who have made such a thing practical. However sophisticated the concept of Vibroseis, it could never have succeeded without the down-to-earth skills of these engineers.

The typical operating pressure of the system is 20 megapascals (MPa) (2900 pounds per square inch, psi), of which about 16 MPa (2300 psi) remains after the pressure loss in the servo-valve. If then the area of the piston is 65 cm^2 (10 in.^{2}), the nominal peak force of the vibrator is about 100 kilonewtons (kN) (23,000 lb-force). Vibrators are normally rated by this nominal-force capacity, which is a crude initial measure of their output. Other things being equal, we obviously like this figure to be high.

But other things may not be equal. In particular, we are much concerned with the frequency response of the vibrator. A typical amplitude-frequency response might appear as in Fig. 6-6. In this (for the moment, and because our geophones are sensitive to particle velocity), we plot the **velocity** of the baseplate as the measure of vibrator output. We remember that particle velocity is the integral of particle acceleration and the differential of particle displacement. On a velocity plot, therefore, a response in which acceleration is indepen-

Fig. 6-5 Main vibrator housing (visible behind the lift ram).

dent of frequency falls at 6 decibels (dB)/octave to the high frequencies, and if displacement is independent of frequency it falls at 6 dB/octave to the low frequencies.

Let us look at the portion of the response between about 40 and 80 Hz in Fig. 6-6. It falls at 6 dB/octave to the high frequencies. It is a response of constant acceleration, and hence of constant force. This is the portion of the response defined by the nominal-force figure discussed above.

Now let us look at the portion between 10 and 40 Hz. Here the velocity is constant. This is the portion where the vibrator is unable to maintain a constant force, because the pump cannot maintain sufficient flow and the servo-valve produces too much drop of pressure.

Now for the very low frequencies. To maintain constant velocity, as the

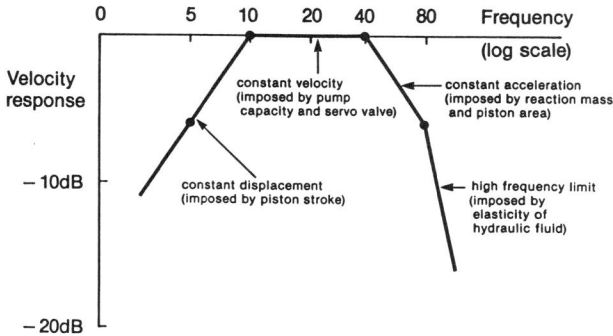

Fig. 6-6 Agencies defining the frequency response of a vibrator. Advancing technology improves some features of this response progressively.

frequency is lowered, requires very large displacements; sooner or later, the piston hits the stops. This limit on stroke (typically 10 cm, or 4 in.) therefore imposes a 6-dB/octave fall at the low frequencies.

Finally, the very high frequencies. In this portion, despite the best we can do in shortening the path between servo-valve and cylinder, we are beaten by the compressibility of the oil. The velocity response falls at 18 dB/octave.

Because of these factors, it is very difficult to make a vibrator that will perform uniformly over a wide frequency range. So we find special-purpose, low-frequency vibrators for seismic penetration to great depths; typically, these have very large reaction masses (7 tonnes) and very large strokes (23 cm, or 9 in.). We find special-purpose, high-frequency vibrators for good resolution at shallow depths; typically, these have very short oilways and advanced servo-valves. And between them we have general-purpose vibrators, which, by adopting reasonable compromises, do their best to give us constant output from perhaps 15 to 75 Hz, or better.

But there is another problem. Besides the frequency response of the vibrator itself, we must consider the frequency response of the coupling between the baseplate and the earth. The ground approximates to a spring. This spring, coupled with the mass of the baseplate (and the earth that moves with the baseplate), yields a resonant system. Damping of the system is provided by the heat of friction and (to a disappointingly small degree) by the radiation of seismic energy. Naturally, we choose our baseplate characteristics to place the benefit of this resonance somewhere in our seismic spectrum. As the springiness of the ground changes (for example, in going from a road to its grass verge, or in crossing a dry river valley), it is inevitable that the frequency response of the vibrator–ground coupling must change (Fig. 6-7).

As always in these matters of resonance, we are between two stools. Do we want good bandwidth or lots of poop? And, of course, we compromise. In the early days of Vibroseis, when the vibrators were short on power, the main concern was to provide a rather sharp peak in the coupling at the frequencies

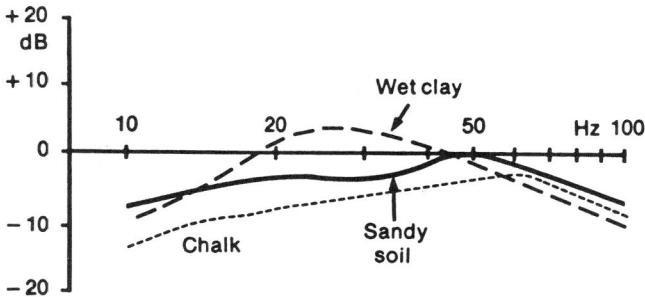

Fig. 6-7   Some typical responses for the vibrator–ground coupling.

giving the best penetration from limited energy. This peak, typically at 25 to 30 Hz, dominated the overall vibrator response. That is why early Vibroseis records tended to look low-frequency and ringy.

Modern vibrators provide a better coupling response. However, it always remains true that there is limited value in improving the frequency response of the vibrator itself if the coupling does not allow the extended frequencies to enter the earth. Thus we must be wary of claims of vibrator response to 300 Hz; does the energy at 300 Hz actually get into the earth?

The preceding discussion, following the historical approach, takes the baseplate velocity as the measure of vibrator output, and considers how this is modified by the coupling resonance. Increasingly, in modern practice, we adopt the view that a better measure of useful output—and one better adapted to dealing with variations of surface coupling—is the **ground force** exerted by the vibrator. We shall return to this in Chapter 7.

We must also consider the phase-frequency response. This proves to be even more important than the amplitude-frequency response because it has a direct effect on reflection time. So large are the phase shifts encountered in a vibrator and its coupling to the ground, and so large are the variations introduced by different ground conditions, that we find it essential to compensate for them in the field. We do this by electronically shifting the phase of the sweep delivered to the vibrator in such a way that the output of the vibrator is maintained locked in phase with the control sweep. This operation is called **phase compensation** or **phase lock,** and it leads us naturally to a discussion of the vibrator electronics.

CHAPTER 7

Vibrator
Electronics

There are several separate functions within the vibrator electronics. One of them is the sweep generator, which we have already discussed. Let us take the phase compensator next.

A basic arrangement, standard until the mid-1980s, is shown in Fig. 7-1. The control sweep comes out of the sweep generator, through the phase shifter (which, let us say, initially does nothing), to the torque motor in the servo-valve. Because of the phase shifts in the vibrator itself and in the vibrator–ground coupling, the particle velocity of the baseplate is out of phase with the control sweep, particularly at the low and high frequencies. This velocity is detected by a geophone (actually an integrated accelerometer) coupled to the baseplate, and its phase is compared to that of the control sweep in the phase comparator. The phase comparison is done by cross-multiplication of the detected signal and the control sweep; after a stage of integration, this has the virtue of being sensitive only to the fundamental in the (distorted) baseplate signal. Then the output of the phase comparator, representing the phase error, drives the phase shifter. The phase shifter changes the phase of the drive to the vibrator, just sufficient to keep the baseplate velocity in phase with the control sweep.

The phase compensator takes a few cycles to pull the two signals into lock; this is minimized by memorizing the last sweep. Once lock is achieved, the velocity of the baseplate is indeed in phase, cycle by cycle, with the control sweep. And, since the control sweeps generated in each vibrator truck are all

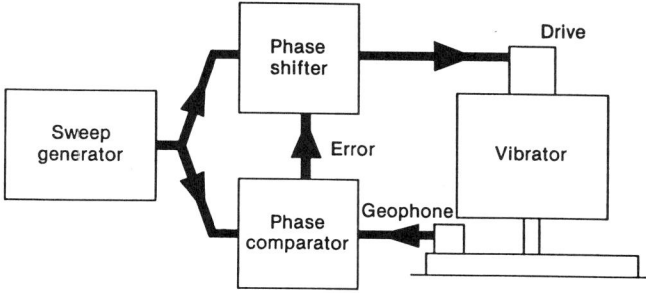

Fig. 7-1 Phase-compensation system.

identical, it follows that all the baseplates are locked together, moving as one. We are safe from the nightmare of having one vibrator push while another one pulls.

But we are not yet sure that our nominal reflection pulse is basically a symmetrical auto-correlation function (Fig. 7-2a). The reason is that, in the system described, the variable locked to the control sweep is baseplate *velocity*. For a spherical wave propagating in a uniform fluid medium, the particle velocity advances phase by 90° within the first wavelength or so from the source. Therefore, to the extent that this conclusion applies to the practical case (a hemispherical wave propagating in a nonuniform solid medium), the natural

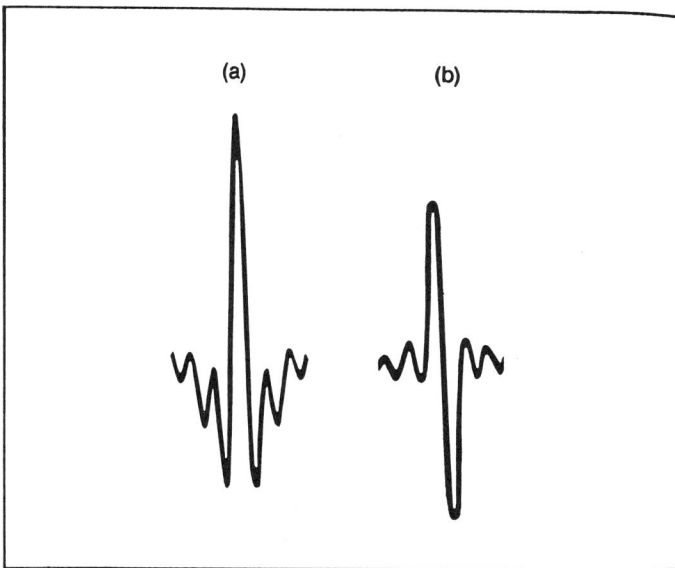

Fig. 7-2 Conversion of the symmetrical auto-correlation (a) to a skew-symmetric form (b).

course of locking the baseplate velocity to the control sweep runs the risk that the reflections will become basically skew-symmetrical (Fig. 7-2b).

As it happens, this effect is largely offset by a practice recommended by the SEG Subcommittee on Polarity Standards. For quite a different reason (a historical one that might not be considered valid today), the recommendation is that the phase of the baseplate velocity should lag the control sweep by 90°. If this is done (and the vibrator electronics have a "90°" setting by which it can be done easily), the 90° advance of particle velocity in the earth is anticipated and compensated.

Even before the mid-1980s, there was not universal agreement on this course. Some operators felt that, in the real earth, we are not sufficiently sure of what happens to particle velocity near the source; they prefer to keep the baseplate velocity in phase with the control sweep (using the "0°" setting). Others observed that the phase-compensation electronics work better if the variable locked to the control signal is the acceleration of the reaction mass, which is a "cleaner" signal. Still others were not concerned with rights and wrongs—only with making new surveys tie with old ones.

By the mid-1980s, however, there was increasing acceptance of the practice of locking the *ground force* to the control sweep. The ground force is taken as a weighted sum of the products of mass and acceleration for the reaction mass and the baseplate. This is computed continuously, and used in place of the baseplate velocity signal in Fig. 7-1. It was also found that on some surface materials it is better to try to keep the ground force constant (with frequency) than to try to keep the particle velocity of the baseplate constant.

We should keep this issue in perspective. We recall that it is in the field that we must pay attention to the *amplitude* spectrum of the signal, because it is in the field that the final signal-to-noise ratio at any given frequency is decided. (If we do not emit enough signal at 37 Hz to overcome the noise at 37 Hz, given all the benefits of stacking and so on, then there is nothing we can do later to rectify this.) Furthermore, it is in the field that we must lock each of the several vibrators to *something,* so that they are all pushing and pulling in phase. However, it is not of the first importance that the *something* should maintain the notional symmetry of the basic Vibroseis wavelet; if it introduces a 90° shift or some other *known* shift, we can always correct that in processing.

The field reports must say whether the phase lock was 0° or 90°, velocity or ground force, because we may need to use this information in the processing.

Back to the vibrator electronics. We have noted that at low frequencies the limit on vibrator output is the stroke. The piston must not be allowed to hit the stops. Accordingly, a transducer senses the relative displacement of piston and reaction mass, and a feedback arrangement decreases the drive to the torque motor as danger approaches.

The servo-valve also has a limit on displacement; therefore, a similar arrangement is provided to prevent excess motion. This represents a limit on oil flow (as does the capacity of the pump). We recall from our discussion of Fig. 6-4 that the stroke limitation controls the performance of the vibrator at low frequencies, while the flow limitation controls it at medium frequencies.

Other functions provided in the vibrator electronics include control of the baseplate lift system, control of the mid-position of the piston, a matched filter for reception of the start-sweep command from the recording truck, remote selection of the sweep from the recording truck, various test facilities, a drive control, and a radio carrier system providing transmission of the sweep and baseplate signals to the recording truck. The last two deserve specific discussion.

The drive control varies the drive applied to the torque motor. There are two things we need to know about this. First, *if we decrease the drive, we change the amplitude-frequency response of the vibrator;* this is because the stroke and flow limits then become effective at lower frequencies. Thus tests of frequency response made at low drive settings are not representative of actual operation. Second, vibrator operators learn very quickly that if they turn down the drive (or the hydraulic pressure) the vibrator gives less trouble. Of course it does. However, for the same ratio of signal to ambient noise, operation at reduced drive slows production dramatically. Somebody in the field has to be vigilant about this.

What we really want is a system that always maintains the largest ground force possible without damaging the vibrator or allowing the baseplate to jump off the ground. Accordingly, the present standard practice is to measure the peak upward force generated, and to compare it first with the static hold-down force provided by the weight of the truck and then with the envelope amplitude of the control sweep; any excess force is then countermanded by turning down the drive. The force is also prevented from rising to the level where the piston would hit the stops. An automatic decrease of drive may also be provided, if we wish, to prevent the total harmonic distortion from rising above a set value.

With so many automatic controls inside the vibrators, we have no direct assurance that the effective output follows the amplitude envelope of the sweep. Modern equipment makes it possible to transmit back to the recording truck the twin measures of applied force and produced motion, from which the true "source signature" of each vibrator can be ascertained.

Whether or not this system is used, we need some scheme for transmission back to the recording truck, in order to test the vibrators. In the jargon, these tests are called *similarities.* On older equipment, similarities require that the vibrators be brought to the recording truck and connected to it by wire; this is practical only once or twice a day. On current equipment the signals (which can be those corresponding to actual production records) are transmitted by radio to the recording truck; there we can make a permanent record of each vibrator

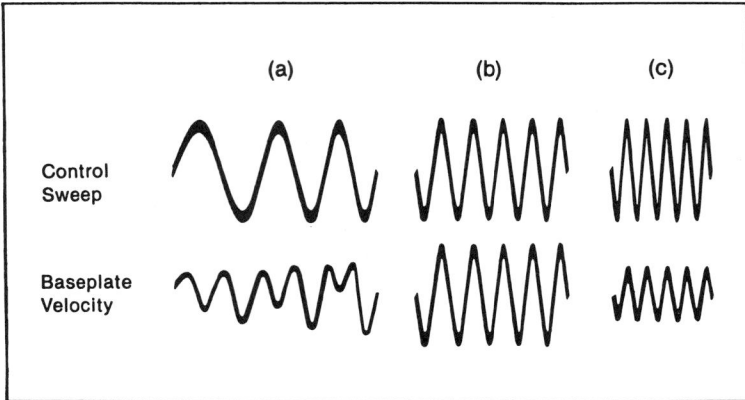

Fig. 7-3   Portions of a similarity test on one vibrator.

on tape, a printout of the phase-lock performance, and perhaps even a test cross-correlation.

Figure 7-3 illustrates portions of an acceptable similarity test at low, middle, and high frequencies. This figure does not use the option, provided in some equipment, of applying automatic gain control to keep the amplitudes constant, or of applying other cosmetic devices. In each trace, the control sweep is the upper of the two traces; it is of constant amplitude and perfect form. At the middle frequencies (b), we see the full amplitude of the baseplate velocity, and good phase lock. At the high frequencies (c), we see some loss of the baseplate velocity; this, we recall, is the effect of the constant-force part of the response. At the low frequencies (a), we see a quite different effect: severe even-order harmonic distortion. The repercussions of this are such that it warrants a section of its own.

CHAPTER **8**

The Effects
of Distortion

Although some distortion is normal in the servo-valve, the main cause of even-order distortion is in the basic nonlinearity of the surface of the earth; it is easier to pull up than to push down. The effect becomes appreciable at low frequencies, where the displacements of the surface become large.

The result is that if we drive the vibrator at 15 Hz the motion of the baseplate may appear to be mostly at 30 Hz. At lower frequencies, we may not even be able to see the fundamental at all.

The effect is most marked if the force applied by the vibrator exceeds the hold-down weight; then the baseplate actually chatters on a hard surface. We remember from the last chapter that modern vibrators minimize this by automatically decreasing the drive when the ground-force measurement approaches the hold-down weight. Even with this feature, however, some even-order distortion is common.

The first point to make about this is that it does not affect the reflection pulse itself. The correlation process, when searching for a reflection at a particular time, does not see the harmonics. This follows, of course, from the extreme selectivity of correlation; the 15-Hz part of the control sweep correlates with the 15-Hz part of the reflected sweep, weak though it may be, while other frequencies yield very little correlation. Thus distortion does not invalidate the reflections themselves. That is not to say we like it; it means that energy has been lost from the fundamental to generate the harmonics, and it increases the practical problems of phase compensation.

A second point, however, is more serious: the distortion destroys the nonrepetitive uniqueness of the sweep; the same frequency can now occur twice. The result is a **harmonic ghost.**

Figure 8-1a shows a single reflected sweep that has suffered distortion at the low frequencies, and Fig. 8-1b shows the control sweep sliding past it at the moment when the reflection is detected. For simplicity of illustration, we ignore frequency-selective effects in the vibrator and in the earth, so the output trace in Fig. 8-1d shows the reflection pulse (e) as a symmetrical auto-correlation function. So far, all is well. But then we note that a spurious partial correlation occurs some time before the true reflection correlation; this is when the second harmonic of a frequency (say 15 Hz) in the reflected sweep correlates with its own frequency (say 30 Hz) in the control sweep (c). The result is a harmonic ghost (f).

Thus, if second-order distortion occurs at the low end of a 15- to 55-Hz sweep of 8-s duration, a harmonic ghost precedes the true event by 3 s.

In practice, the distortion occurs over a *range* of low frequencies, gradually becoming less serious as the frequency rises. The harmonic ghost is therefore a diffuse event, containing the harmonics of this range of low frequencies.

The danger from harmonic ghosts, obviously, is that they do not correspond to arrivals at the time indicated. They have the moveout of real events at some other time. Furthermore, they may obscure genuine reflections. Therefore, we must reduce harmonic ghosts to some negligible level.

This leads us to a discussion of upsweeps and downsweeps. With an upsweep, as illustrated in Fig. 8-1, the harmonic ghost *precedes* the true event.

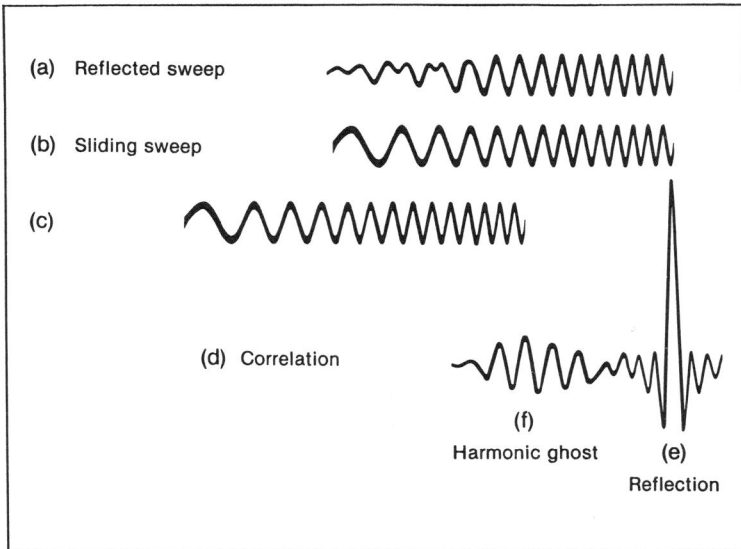

Fig. 8-1 Generation of harmonic ghosts by an upsweep.

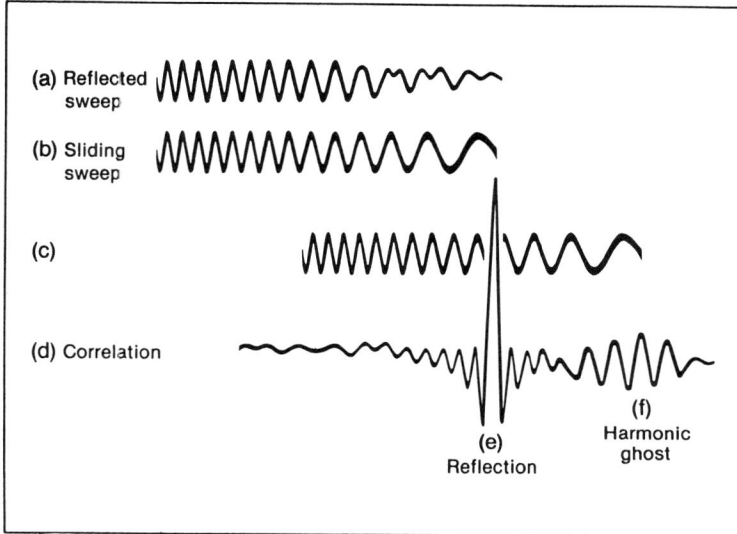

Fig. 8-2   Generation of harmonic ghosts by a downsweep.

For a downsweep, as illustrated in Fig. 8-2, the harmonic ghost *follows* the true event. Which is the better choice?

Clearly, the upsweep. For if the ghost comes in 2 s *before* an event, simple geometrical divergence decreases the chances that the ghost may obscure a true reflection. But if the ghost comes in 2 s *after* an event (Fig. 8-3), its ghost is much more dangerous. That is why all our discussion so far has been in terms of upsweeps.

However, there are other factors in the choice:

- We have noted that the phase compensator may take a few cycles to lock. A few cycles is much less of the sweep if we use a downsweep. As mentioned earlier, this disadvantage of the upsweep can be reduced by memorizing the low-frequency phase shift from one sweep to the next, so that the phase compensator is in or near lock at the start of each sweep (after the first).
- The hydraulic accumulators make their main contribution at the low frequencies, when extra oil flow is demanded. A downsweep means that the accumulators are full during most of the sweep; they give up their charge only at the end.
- Many observers make their judgment on the basis of the tingle-toe test, and this seems to lead them to prefer downsweeps. The preference may be explained by the accumulators, or it may be a feature of compaction of the soil; it is well established that a sweep on virgin soil is much less effective than on compacted soil, and this may possibly favor the downsweep.

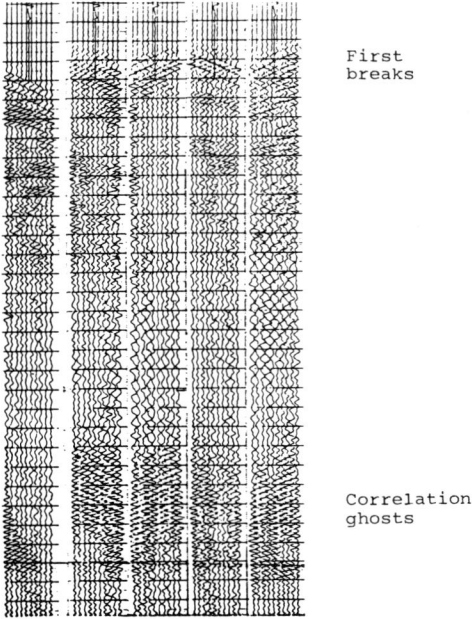

First
breaks

Correlation
ghosts

Fig. 8-3 Correlation ghosts, with the alignment of the first breaks, evident on field monitor records.

The choice between upsweep and downsweep was hotly argued in times past. Today, however, we can relax the argument for upsweeps, because longer sweeps now allow the ghost time to be more than the length of the record. Only if we are trying for a very large bandwidth are we apprehensive about harmonic ghosts. However, we always make a point of checking the similarities to see at what frequencies the distortion becomes appreciable, and then of calculating the ghost time.

CHAPTER 9

Instrumental
Considerations

The recording instruments used with Vibroseis are standard. In fact, they do not really need to be as good as this, because the signal-to-noise benefit of correlation applies to instrument noise as much as it does to ambient noise. Furthermore, the dynamic range of the geophone signal is less with Vibroseis than with explosive work because of the offset (and, in older work, because of the long arrays). However, we normally apply an excess of caution, and use standard instruments.

With Vibroseis, there is a neat way of eliminating the phase response of the recording instruments. This is to pass the control sweep through a standard recording channel on its way from the sweep generator to tape (or to the correlator). Since the control sweep and the geophone signals have then been through the same phase response, this phase is eliminated from their cross-correlation. A separate compensation for the phase response of the geophone is still desirable, however; this may be applied during processing.

Some crews record the output of the vertical stacker on tape as the final output from the field; others save tape and processing time by performing the correlation in the field. In this case, with modern fast correlators, we have the option of correlating either before or after vertical stack. The argument for performing the vertical stack first is that it allows sophisticated suppression of some types of noise; we shall discuss this later. The argument for performing the correlation first is that it allows different sweeps to be used within one source point; one application of this is discussed in item 8 of Appendix A.

Whether or not we do the final correlation in the field, we must provide some rudimentary correlation for the observer. Without this, the observer really has no way of judging whether the field technique is satisfactory. For these purposes, the monitor correlator can be surprisingly primitive; single-bit correlation, which looks only for coincidence of sign between the received signal and the transmitted sweep, is often visually indistinguishable from correlation with full numerical significance.

For the observer's purposes, it is desirable that the signal against which the monitor correlation is made can be a *modified* version of the control sweep. In illustrating why this is so, we can find several additional and interesting features:

- We consider a field operation using 12 sweeps per SP, 15 to 75 Hz over 24 s with a 30-s cycle, and ask: Is there any real benefit from the 60- to 75-Hz range? So the observer performs the monitor correlation against a sweep of 15 to 60 Hz over 18 s; this matches the transmitted sweep exactly, cycle for cycle, except that it omits 60 to 75 Hz. The result is the record that would have been obtained with a 15- to 60-Hz sweep; the test does not require the actual transmission of a 15- to 60-Hz sweep.

- Perhaps, even after the best that can be done with arrays, the records show serious ground roll concentrated between 15 and 20 Hz. The observer makes two records of the same SP: one with a 24-s sweep of 15 to 75 Hz, and one with a 22-s sweep of 20 to 75 Hz (that is, again with the same sweep rate). The observer correlates them both against the 20- to 75-Hz sweep. If the records are equally good, the observer can continue using the 15- to 75-Hz sweep; the processing can handle the ground roll. If the record from the 20- to 75-Hz sweep is superior, the observer concludes that it is better not to excite the ground roll than to excite it and filter it out—a nonlinear situation. The problem is probably harmonic distortion in the geophones or in their coupling with the ground; it cannot be solved in the processing. A severe problem would indicate that the observer should change to a 20- to 75-Hz sweep for production; this, of course, requires changing the sweep rate to maintain a 24-s sweep.

These examples illustrate some of the intriguing features of correlation viewed as a filter. This is explored further in Appendix A (items 5 to 12).

Our final instrumental topic must be **sign-bit recording.** In this the entire instrumental chain (not just the correlation) operates with a single bit, representing only the sign of the signal. Sign-bit recording can be considered for any system in which the reflected signal is ordinarily smaller than the ambient noise. Vibroseis, clearly, is such a system. The simplicity of sign-bit recording allows many more channels, and this in turn facilitates economic three-dimensional surveys.

When we accept sign-bit recording, we do so *on balance;* we say that the economic benefit of more channels outweighs our concern that we may lose some confidence in the amplitude, character, and frequency content of our reflections.

In Vibroseis using sign-bit recording, then, the signal from each geophone array is sent to the vertical stacker as a sign-bit sequence (that is, as a sequence that says only whether the geophone signal is positive or negative). The information content is present only as a modifier of the time of the zero-crossings of the noise. As the vertical stacking proceeds, the signal acquires some amplitude contrasts; from the stacker onward, these amplitude variations are preserved normally.

CHAPTER 10

Practical
Considerations
in the Field

Many of the factors deciding the field technique are the same for Vibroseis as for the impulsive sources. Thus the distance from the source to the far geophone group (the **far-group offset**) usually approximates the depth to the main target, as in impulsive work. The distance to the near geophone group (the **near-group offset**) usually does not exceed the depth of the shallowest reflector of interest, and is preferably less if the source-generated noise allows. The distance between groups is ideally chosen to avoid "spatial aliasing"; this means a group interval not exceeding $V/2f_m \sin \theta$, where V is the effective seismic velocity, f_m is the maximum frequency maintainable through the earth and the processing, and θ is the maximum reflector dip. These considerations therefore define the spread geometry and the number of recording channels, just as in impulsive work.

With a surface source such as Vibroseis, however, we must pay more attention to source-generated noise than we would do with dynamite in deep holes; as suggested in Fig. 10-1, the vertical motion of the baseplate generates surface waves directly. In the past, this concern led to long geophone arrays at each group and long source arrays at each source point. Such long arrays inevitably caused the loss of high frequencies on shallow reflections at long offset and on dipping reflections. For anything but coarse reconnaissance work, this loss is harmful; in our search for crisp reflection pulses, we must minimize any avoidable loss of the high frequencies. These considerations have led to the

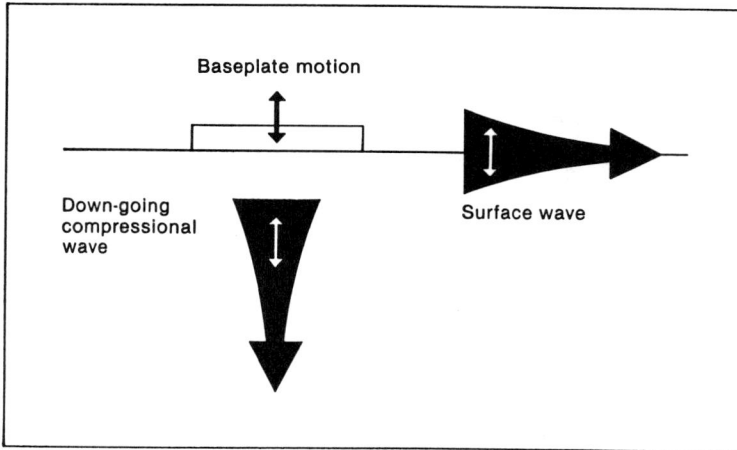

Fig. 10-1 Reminder of the generation of surface waves by a surface source.

stack-array approach to array design, in which the arrays in the field are of modest length (usually equal to the group interval) and in which the long array necessary for the attenuation of source-generated noise is formed in the stack.

Figure 10-2a shows the reflection paths for a sixfold common-midpoint stack, obtained with a split spread. (The figure suggests a buried explosive source, but is equally applicable to Vibroseis.) The array length is taken as substantially equal to the group interval, so on the ground there is an even and continuous sequence of geophones all along the spread. When the resulting recordings are gathered into the common-midpoint arrangement of the figure, we can see the way in which the source-to-group offset varies within the gather. This is plotted in Fig. 10-2b; we see that for the particular spread geometry of Fig. 10-2a there is an even and continuous sequence of geophones all across the **gather,** as well as all along the **spread.**

In this case, the process of stacking forms one long array the whole length of the spread. The length of this array is sufficient to attenuate source-generated noise of all wavelengths observed in practice, and so in principle the stack-array approach provides virtually complete suppression of this noise. In practice, of course, the suppression is less than total; variations of elevation along the line, and variations of surface material from source point to source point, reduce the degree of suppression obtained. However, where the degree of suppression obtainable is sufficient, the stack-array approach to array design is to be preferred over the traditional approach using long arrays; this is particularly so in detail surveys, where the smearing and high-frequency loss imposed by long individual arrays is so harmful.

The stack-array requirement that the individual arrays should be continuous and end-to-end in the gather is not satisfied merely by having them

(a)

(b)

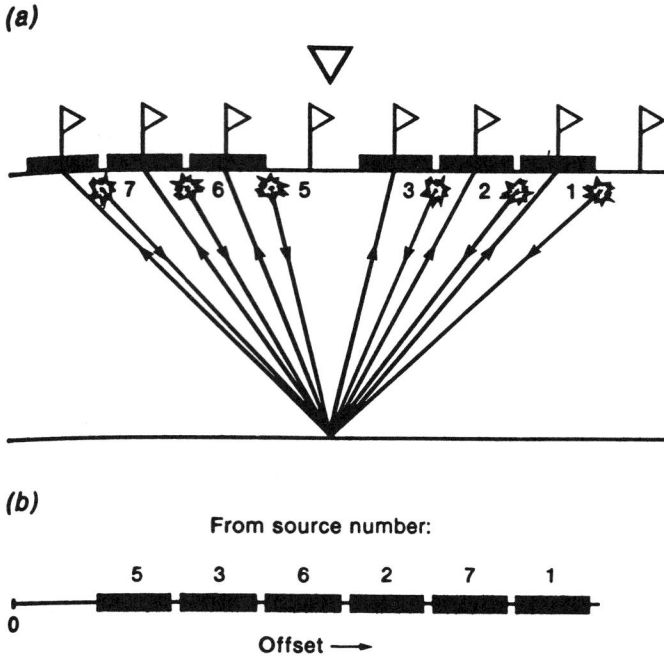

Fig. 10-2 Physical arrangement of a sixfold common-midpoint gather (a), and the formation of the stack array from individual arrays at continuously increasing offset (b).

continuous and end to end on the ground. *Indeed, for Vibroseis work on land we are led to one preferred arrangement:*

- A split spread
- Group length equal to group interval
- Source interval equal to group interval
- Effective source point halfway between group centers

This highly practical arrangement conforms to the example we used in Fig. 5-4 (reproduced as Fig. 10-3). As in the figure, we use the flags as the group centers; then we lay each group symmetrically about the flag, and the vibrators move to build up a source array from flag to flag. (Alternatively, some crews prefer to lay the groups from flag to flag, and to use the flags as the source-array centers.)

Basic to the stack-array approach, then, is a continuous and even spacing of geophones all along the spread, and a continuous and even spacing of vibrator positions all along the line. This arrangement leads to a fold of coverage equal to half the number of the geophone groups; thus 60 channels (30

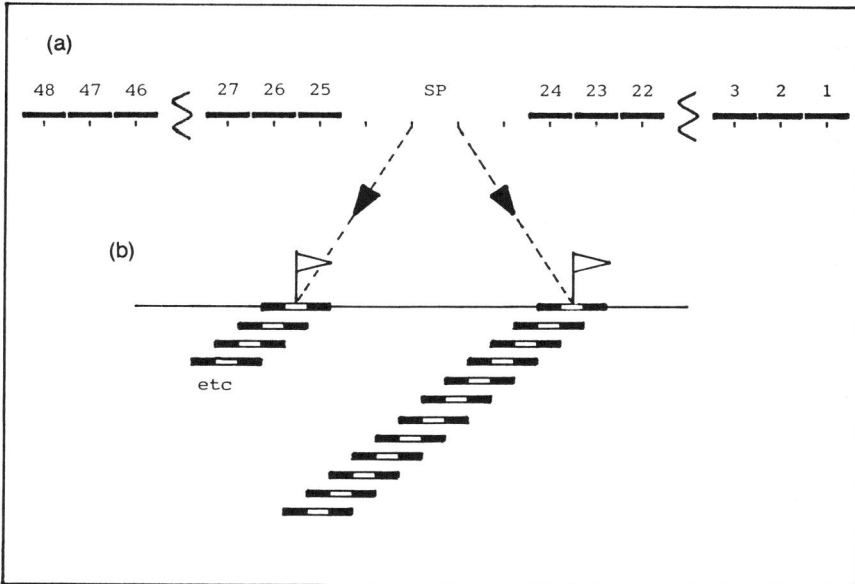

Fig. 10-3 Reminder (from Fig. 5-4) of a practical arrangement for the source array, centered on the midpoint between geophone stations.

groups on each side of the source) yield 30-fold stack, 96 channels yield 48-fold, and so on.

This arrangement contains some exposure to spatial aliasing if we erred in estimating the maximum dip θ. To avoid this, some geophysicists choose to *remove* any reflections with greater dips. This can be done, with the arrangement described above, by using a group length equal to *double* the group interval. Some crews following this philosophy lay groups of this length on the ground; this involves overlapping (and the risk of tangling) of the geophone strings. Others keep the group length equal to the group interval in the field, but effectively double the length by "mixing" the recordings from adjacent groups during processing; this is more convenient.

Both methods have the advantage of satisfying the stack-array continuity on both sides of the spread independently, and thus require less constancy of the source-generated noise from SP to SP.

The course recommended here is the basic technique outlined above, with the group length equal to the group interval. If it is found in processing that dips are greater than those allowed for in the choice of group interval, or that the source-generated noise is very different from SP to SP, then the mixing approach can be used; otherwise, all the benefits of short arrays are preserved.

The stack-array criterion, then, is that the geophone groups should be continuous and end-to-end across the gather. This criterion is satisfied by either arrangement described above. However, the criterion is actually broader than this; for a split spread shot between the groups, it is still satisfied if *either* the geophone-array length *or* the source-array length is the required integral number of group intervals. This rather surprising conclusion allows a welcome flexibility in Vibroseis operations. Where the stack-array is sufficient to handle the source-generated noise by itself, we can dispense with the source array; thus we can keep the vibrators stationary, either right at the SP, nose to tail, or conveniently spaced apart. In uneven desert country, where it is practical to space the vibrators apart laterally, we can use any convenient *areal* array. And if the source-generated noise is *not* handled adequately by the stack array, we can bow to traditional practice by lengthening *one* set of arrays, either at the source or at the geophone, while keeping the other set satisfying the stack-array criterion.

We see, then, how the spread geometry is defined. The target depth defines the far-group offset. The shallowest reflector of interest, coupled with the intensity of the source-generated noise, defines the near-group offset. The degree of lateral detail required for the target, coupled with the spatial-aliasing constraint, defines the group interval. The group interval, coupled with the stack-array criterion, then defines the group length, the source interval, and the fold of cover.

We turn now to the choice of sweep frequencies. We must choose the low-frequency limit and the high-frequency limit.

Occasionally, when the source-generated noise is very troublesome at the low frequencies, we take advantage of a particularity of Vibroseis, and choose the low-frequency limit *above* these frequencies. We then solve the problem at its root, by not generating the troublesome surface waves.

But when this is not necessary, we prefer to keep the low frequencies in the sweep. A significant improvement in octave bandwidth (and therefore in the "stand-out" of the reflections) can be realized very cheaply at the low frequencies; it takes very little sweep time to extend the sweep from 14 down to 10 Hz, and this is another half-octave of bandwidth.

The upper limit of the sweep is more difficult. To extend the sweep from 56 to 80 Hz is also half an octave, but this reduces our production by about 30%. We are much more concerned, therefore, to ask whether the vibrator response is adequate for the extra bandwidth, and whether the earth will pass it at any recoverable level; if the response at target level is down 30 dB at 80 Hz, we are unlikely to get much benefit from a linear sweep to 80 Hz.

The choice of upper sweep frequency is therefore a compromise between bandwidth and cost. The difficulty in the choice, as we noted before, is that the decision made in the field is final; we cannot receive frequencies we do not send. Therefore, we take records with different upper frequencies and compare them (particularly on the near half of the spread and at the earliest reflection time of interest); the sweep limit is taken somewhat higher than the frequency beyond which there is no visible change in reflection character.

The number and length of the sweeps are also established by experiment. Here we should remind ourselves that the final ratio of signal to ambient noise is decided by the *total sweep energy radiated per kilometer or per mile* (in the useful frequency band, of course); our concern is to emit the necessary energy in the minimum time and so to maximize the production.

What should we do if a vibrator gets sick? By how much do we increase the number and/or duration of the sweeps?

To a first approximation, a decrease from four vibrators to three reduces the signal-to-noise ratio in inverse proportion, to three-quarters. To recover the original signal-to-noise ratio, we must increase the total sweeping time by $(4/3)^2$; this is because the signal increases by a factor of $(4/3)^2$, and the noise (which increases as the square root of the time) by $4/3$. Therefore, contracts that specify four vibrators in the field often allow temporary operation with three vibrators, provided that the number of sweeps (or the sweep duration) is increased by 80%. If a second vibrator fails, the required number of sweeps would be four times the original number, and so would represent a very large loss of production; therefore, the contract usually forbids operation when two vibrators are sick. In areas requiring less vibrator effort, the contract may specify only three vibrators in the field; then, when one of these fails, the number (or duration) of sweeps should ideally be increased by $(3/2)^2$, or $2\frac{1}{4}$. Again this is a heavy burden of sweeping time.

In the interests of production, many supervisors allow these rules to be bent on days when the ambient noise is small. But they also press for good maintenance and rapid repair of the vibrators; because the above rules depend on the statistics of the noise, many vibrators always give more comfort than many sweeps.

In all the foregoing material we have considered the noise to be the usual ambient noise (wind, cattle, distant storms, and so on) that we encounter in explosive work. But we should also consider traffic noise. In explosive work we seldom work along roads, and when we do so we can usually wait for a quiet period in the traffic noise before shooting. Some Vibroseis operations, however, are conducted exclusively along roads, and traffic may be the major source of noise. Then, since the recording is continuous, we cannot rely on quiet periods. For these operations, important noise-abatement techniques have been developed, taking advantage of the special characteristics of some types of noise.

First let us consider the effect of one burst of **impulsive** noise (for example, a flash of lightning) during a Vibroseis sweep. The signal entering the recording instruments might appear as in Fig. 10-4a. We see the usual picture of ambient noise, below which (we hope) are submerged many good reflected sweeps. Then we see one enormous burst of impulsive noise, at an amplitude many times that of the ambient noise. Clearly, *we know it is noise;* we cannot visualize any circumstance that would make the reflected sweep signal (or even the surface waves) increase dramatically in amplitude *for a short time.*

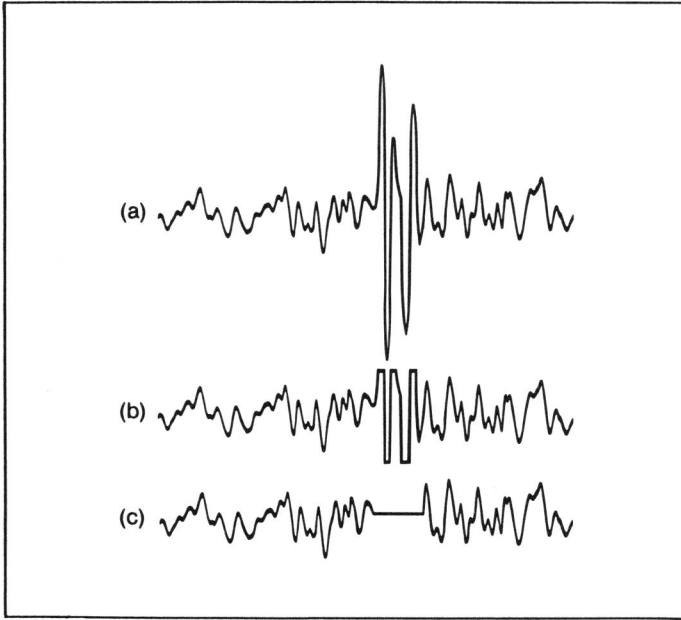

Fig. 10-4 Clipping or zeroing of impulsive noise before correlation.

We have two easy choices: we can **clip** the impulsive noise at some level safely above the usual signal (Fig. 10-4b) or we can zero it (Fig. 10-4c). Let us suppose that we zero it, say for 100 milliseconds (ms). Then each reflected sweep is lacking a small fraction of its total length. The effect of this on our ability to detect the reflections by cross-correlation is quite negligible. Vibroseis is almost immune to a short period of high-amplitude noise, provided that the noise can be recognized as such and *be clipped or zeroed accordingly.*

However, we should note that, in addition to making a miniscule change of reflection amplitude to most or all of the reflections, the zeroing has cut a frequency notch in them. What is more, this notch is at a different frequency for reflections of different time.

If the lightning flashes several times during a recording, we can do the same thing. Although we now begin to be less sure of the fidelity of reflection amplitude and character, the clipping or zeroing of the noise bursts clearly remains the best thing to do.

But if there are many flashes? Again we do the same thing. Now we are hoping that the many sweep cycles that enter one vertical stack are notched in randomly different places, so that a total notch at one or more frequencies is unlikely. But one thing is certain: as the amount of zeroing becomes appreciable, we must start to emit more sweeps. More noise needs more signal.

So much for impulsive noise. It is undesirable, but its characteristics of high amplitude and short duration tell us what to do, and what we do is easy.

Traffic noise, however, is different. It comes in packets of several seconds; its amplitude may be very large, or it may be comparable to the ambient noise. What are we to do?

The first step, normally, is to assume that any *unusually* large amplitudes are caused by traffic noise. During the actual recording, we have no other criterion for recognizing traffic noise. Then we automatically turn down the gain during the noisy periods, and turn it up again during the quiet periods, to keep the general amplitude level approximately constant. The content of genuine reflection signal is thus low during noisy periods and high during quiet periods.

The method of doing this is illustrated in Fig. 10-5. Figure 10-5a is the incoming waveform from the geophone. It is divided into windows of length less than the typical duration of a packet of traffic noise; thus the length of each window might be 0.25 to 0.5 s. The mean amplitude is calculated for each window, and ascribed to the center of the window as in Fig. 10-5b. A gain function is calculated proportional to the reciprocal of these mean amplitudes, with linear interpolation between them (Fig. 10-5c). The original waveform is then multiplied by this gain function, sample by sample, to yield Fig. 10-5d. The modified waveform, now normalized to approximately constant amplitude, proceeds into the vertical stacker.

The vertical stack of many such records has an improved signal-to-noise ratio. However, as with the zeroing of impulsive noise, we have lost some confidence in the amplitude and frequency content of our reflections. So we ask: Can we memorize what we have done, and in some way compensate any

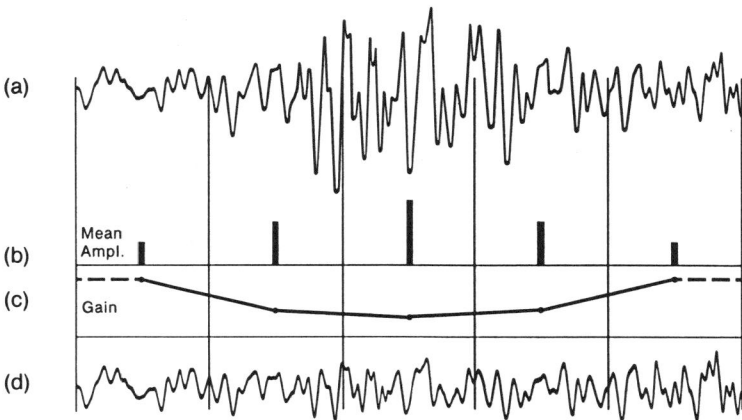

Fig. 10-5   Burst of traffice noise and its treatment by turning down the gain.

harmful effect *after* we have achieved the benefit of vertical stacking? One interesting technique for doing this is **diversity stacking,** illustrated in Fig. 10-6.

Figure 10-6a suggests two successive windows of the first record; both, we shall say, have an average amplitude of 1. Figure 10-6b suggests the same two windows of the second record; the first has the same average amplitude of 1, but the second has a burst of traffic noise that raises the average amplitude to 10. Figure 10-6c represents a normal vertical stack; the two traces are added and the sum divided by 2. The second window is still dominated by noise.

Figures 10-6d and e represent the same two records; however, in each window the record is multiplied by the reciprocal of the *square* of the average amplitude. The two scaling factors are 1 and 1 for record 1, and 1 and $1/100$ for record 2. Record 1 remains unchanged in both windows. Record 2 is unchanged in the first window, but in the second it is now very weak instead of being very strong. In Fig. 10-6f we add the two scaled records. Then in each window we divide by the sum of the two scaling factors, which is 2 in the first window and approximately 1 in the second. The result is the diversity stack of Fig. 10-6g, which has the desired amplitude of 1 in each window.

We can see what the diversity stack has done: in the first window it has used both records equally, while in the second, virtually, it has used only the quiet record. This is clearly the best thing to do. And it has done it automatically.

If the noise on record 2 had been larger than 10, the diversity stack would have used even less of it. If it had been less than 10, the diversity stack would have used rather more. Again, this is clearly the correct course.

In Fig. 10-6 we have added the low-amplitude records as if they were

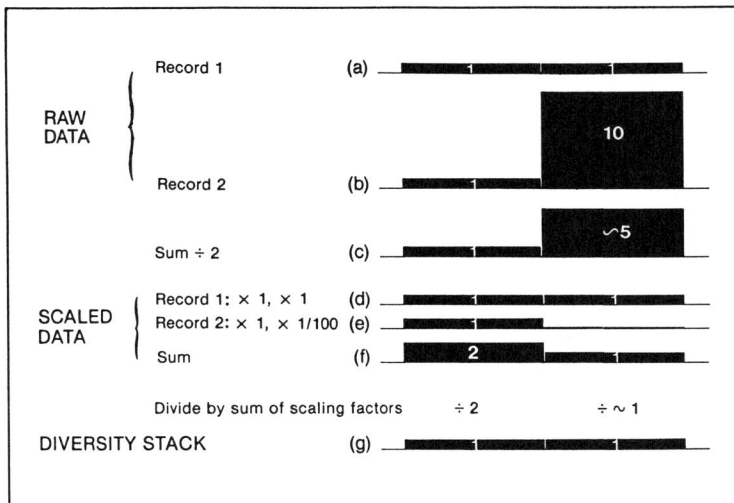

Fig. 10-6 Principle of diversity stacking.

actually *signal*. We recall that, in fact, a good Vibroseis record is usually dominated by ambient noise (as distinct from traffic noise). This changes the numbers of the figure somewhat, but it does not change the generally beneficial action of the diversity stack; the traffic noise remains effectively attenuated.

What is more, the diversity stack allows us reasonable confidence in the amplitude and frequency content of the reflections. Of course, we have to protect it against unusually weak sweeps (the vibrators on peat) and dead channels; otherwise, these would dominate the stack. And we cannot ask the impossible; it can do nothing about a uniform high level of noise. But the diversity stack remains a very *smart* process.

The diversity stack works best with a large fold of vertical stack—many records per source point. This would lead us to use many short sweeps, rather than a few long ones. But there is another consideration, also imposed by traffic noise. Figure 10-7 depicts by a heavy line the frequency–time relation for a long sweep (for example, our previous illustration of a 24-s sweep, with both the move time and the record length equal to 6 s). The heavy parallelogram is a frequency–time map showing what frequencies are used in the correlation as a function of time. Noise inside the parallelogram is noise; noise outside it is irrelevant.

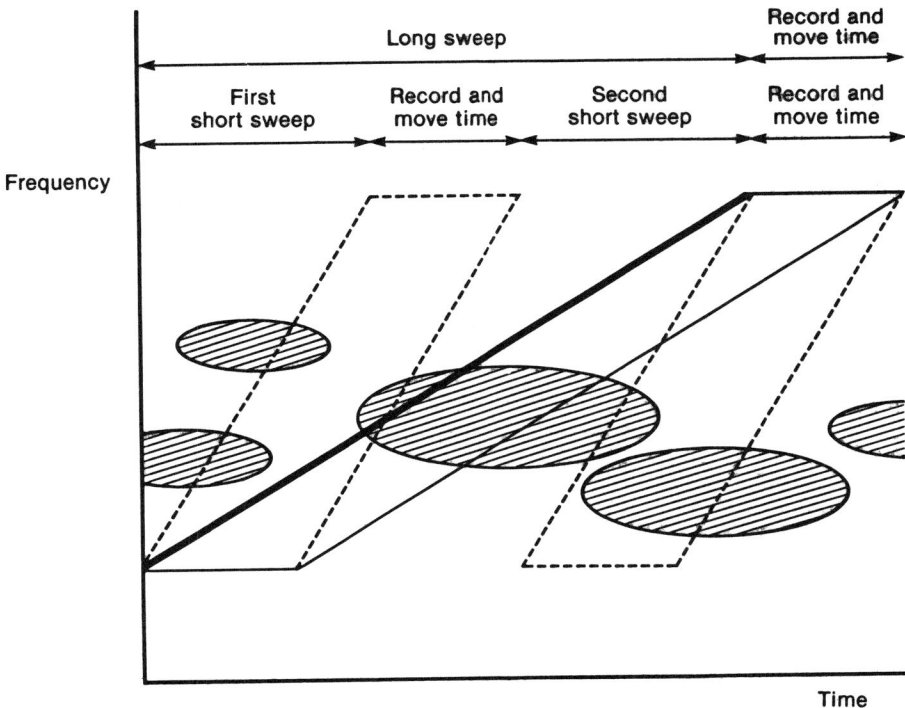

Fig. 10-7 Mapping of packets of traffic noise on a frequency–time plot.

Also suggested in the figure are packets of traffic noise, each a few seconds in length, each contained in a fairly narrow frequency band dependent on vehicle type and speed. Only the packets or part-packets contained within the parallelogram are harmful.

If, instead of the single long cycle of the heavy parallelogram, we use two cycles of half the length, the system becomes susceptible to noise within the two dashed parallelograms. Statistically, this must be worse.

Repeatedly, in the foregoing material, we have discussed the choice between many short sweeps or fewer long ones. Now we can summarize. The arguments for many short sweeps are:

- Diversity stacking works best with a large fold of stack.
- Many vibrator positions provide a smooth source array.

The arguments for fewer long sweeps are:

- Production is increased by the minimization of move time.
- Long sweeps allow greater bandwidth without generating harmonic ghosts.
- Where the vibrators are working on loose soil, compaction is improved.
- Long sweeps are less susceptible to bursts of traffic noise.

A typical compromise would be the use of 8 sweeps per source point. If three or four vibrators are spaced out across the source array, the array then has 24 or 32 elements (which is ample for a smooth response) and the vibrator move between sweeps is practical. Furthermore, the diversity stack functions acceptably well. In particularly noisy areas, 12 or even 16 sweeps are used.

CHAPTER 11

The Problem
of
Optimizing Resolution

As the large petroleum-bearing structures are found, the search turns to traps of progressively smaller areal extent and to reservoirs of progressively smaller thickness. We call this the search for better **resolution.** It has two aspects: the search for **lateral resolution,** which is concerned with features of small lateral extent (particularly faults, buried river channels, bars, and reefs) and **vertical resolution,** which is concerned with thin layers.

In fact, the two aspects are interrelated, and to achieve good lateral resolution it is important to have good vertical resolution also. Since it is in the search for vertical resolution that Vibroseis differs fundamentally from the other methods, we shall concentrate on vertical resolution.

To achieve vertical resolution we must be able to distinguish as separate events the reflections from the top and base of a thin layer. Obviously, this requires short, sharp reflections. Each reflection, we remember, is the convolution of the auto-correlation function of the Vibroseis sweep with the earth filter. The first essential, then, is a short sharp auto-correlation function. Figure 4-5 (reproduced as Fig. 11-1) reminds us that a short, sharp auto-correlation function means a sweep of large bandwidth. Therefore, we can restate the objective of vertical resolution as that of transmitting a sweep of large bandwidth, and of preserving this bandwidth through the earth, the instruments, and the processing.

Generating a sweep of large bandwidth is easy, of course. The problem lies in getting that bandwidth into and through the earth, and through the

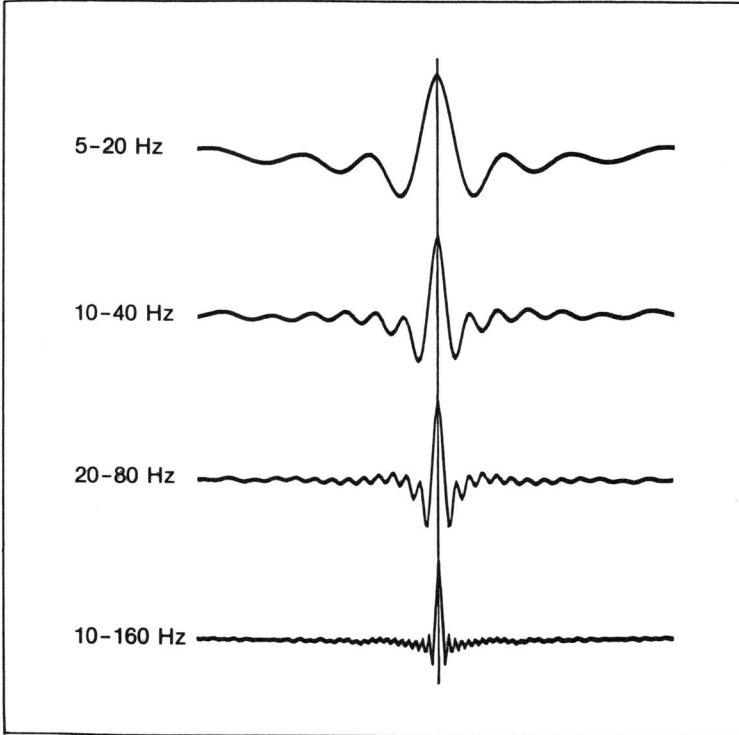

Fig. 11-1 Reminder (from Fig. 4-5) that a short, sharp auto-correlation function requires a sweep of large bandwidth.

recording and processing system. Figure 11-2 identifies the agencies that oppose us in this effort in the field. Figure 11-2a suggests a typical peaked response for the combination of the vibrator and its coupling with the ground. Figure 11-2b shows the likely loss of the high frequencies occurring in the earth filter; it is progressively greater on late reflections than on early ones. Figure 11-2c reminds us that the geophone is a low-cut filter. Figure 11-2d shows a typical response for a long geophone or source array (at a given offset, reflection time, and dip). Figure 11-2e shows the response of the recording instruments; a low-cut filter is imposed to assist the suppression of source-generated noise, and a high-cut filter to allow safe sampling and digitization of the output. Figure 11-2f represents the overall effect of all these agencies; the response is quite sharply peaked, with a modest low-frequency droop and a steeper high-frequency droop worsening with reflection time.

In the early days of Vibroseis, when the vibrators were not as good as they are today, this peaked response led to the use of quite narrow-band sweeps. The sweeps were chosen to straddle the peak of the observed response; 1½ octaves was a typical bandwidth, perhaps from 14 to 40 Hz. There was no point in

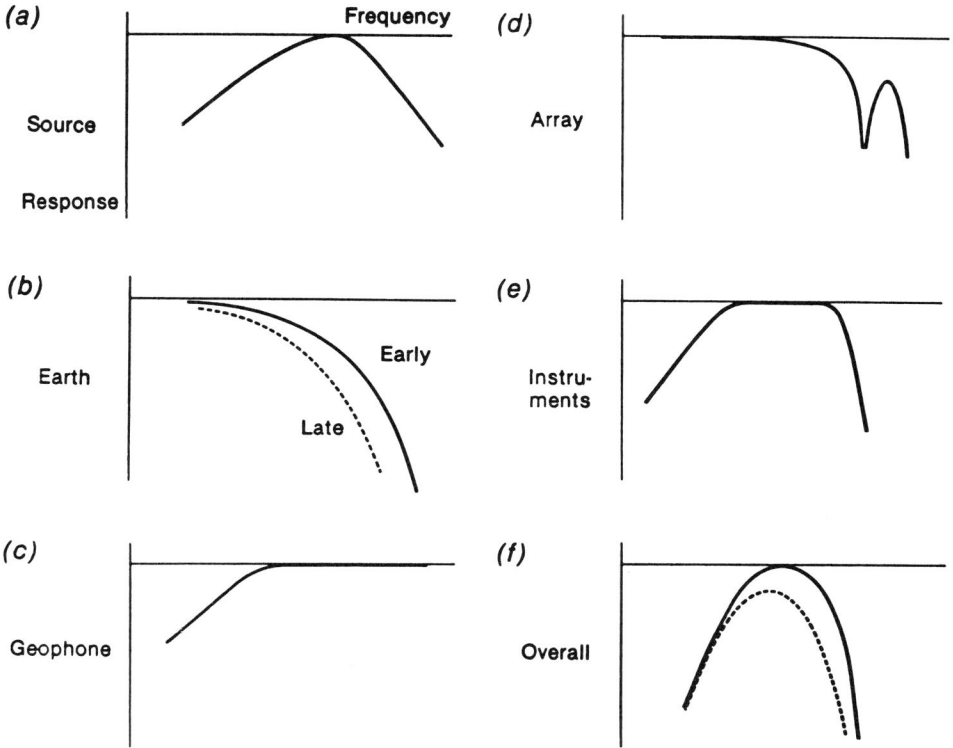

Fig. 11-2   Agencies that limit the reflection bandwidth.

wasting time sweeping frequencies that were hopelessly attenuated in the system and the earth. In those days, therefore, Vibroseis was characterized as a low-resolution system.

Today we can do better. But we still have to fight the earth and the system, and it is still true that there is no point in emitting frequencies that are attenuated beyond hope of recovery. Therefore, for general-purpose work, we choose the sweep frequencies according to the rules set out in Chapter 10, balancing bandwidth and cost. For the earth-and-system response of Fig. 11-2e, the chosen sweep limits might be as shown in Fig. 11-3. The dark-gray limit at the high frequencies would be appropriate if the target is shallow (full line) and the light-gray limit if the target is deep (dashed line). The low-frequency limit is sometimes dictated, as we also agreed in Chapter 10, by the wish to avoid generating excessive surface waves. If this is not a consideration, or we have a better solution to the problem of surface waves, we tend to set the low-frequency limit low; we recall that this is because it is easy and inexpensive to add extra octave bandwidth at the low frequencies, and so we have very little to lose by being generous with these frequencies.

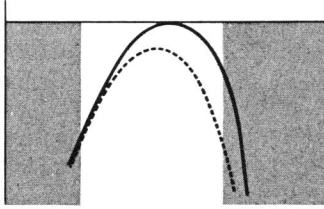

Fig. 11-3 Sweep limits that we might set in general-purpose work for the case of Fig. 11-2f.

Let us now see how this general-purpose practice changes when our concern is to optimize the vertical resolution. As the first step, we must clarify in our minds how much of the effort must take place *in the field* and how much can be effected later, *in the processing*.

11.1 DECONVOLUTION AND SIGNAL-TO-NOISE BANDWIDTH

For the moment, let us suppose that we send a sweep of *infinite* bandwidth to the vibrators. Then the spectrum of the correlated reflections corresponds exactly to the appropriate overall response of Fig. 11-2e. This is reproduced as Fig. 11-4b. The reflection pulses themselves have a corresponding narrow-band form, such as that of Fig. 11-4a. Then we may take these reflection pulses into the computer, and apply the process of **deconvolution.** In effect, this process measures the spectral component a at frequency f in Fig. 11-4b, and computes a multiplier m to bring this value to a new desired value $d = am$ (Fig. 11-4c and d). The deconvolution process therefore has a response *inverse* to the reflection spectrum, and the output of the process has a *flat* reflection spectrum. The output pulse has much larger bandwidth than the input pulse, and hence a sharper form (Fig. 11-4e).

At first sight, the deconvolution process appears to be the complete answer to the problem of the losses in the earth filter and the system. It is not. Figure 11-4 is unrealistic, because it contains no noise. In Fig. 11-5 we see a more realistic situation; the signal spectrum has its usual peaked form, and the signal is superimposed on a background of noise. The content of the noise is shown as falling to the high frequencies; this would be typical for some forms of ambient noise. The signal exceeds the noise only over a certain band of frequencies. We define the signal bandwidth in the usual way (at the points that are so-far down from the peak), but *we define the signal-to-noise bandwidth as that band over which the signal exceeds the noise.*

Now let us consider a common case where the signal has difficulty rising above the noise (Fig. 11-6a). This might be, for example, early on a single field record, or deep in the section on stacked data. The shaded region, from f_1 to f_2, is that in which the signal exceeds the noise. The spectrum of the *combination* of signal and noise—the input seen by a statistical deconvolution process—is shown in Fig. 11-6b. The deconvolution operator (Fig. 11-6c) is calculated to flatten this spectrum; it is therefore inverse to the combination of signal and

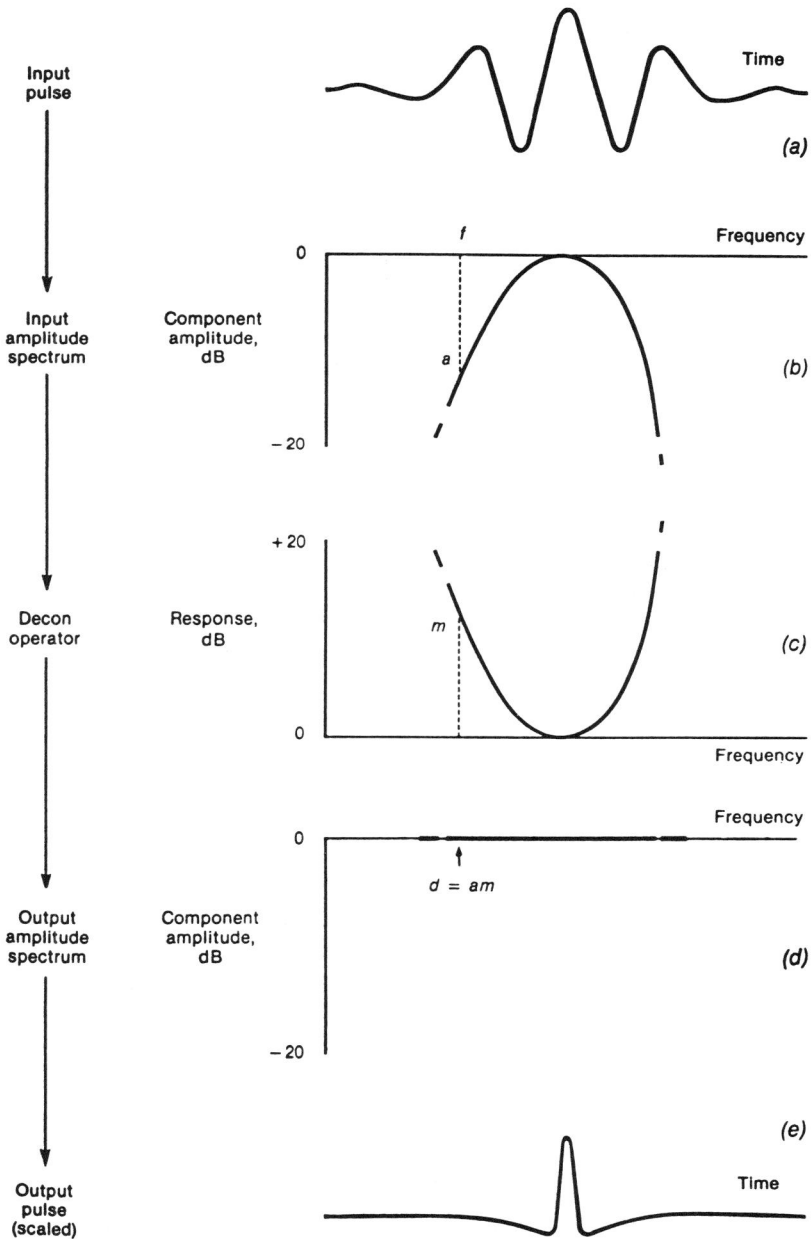

Fig. 11-4   Action of deconvolution in whitening the spectrum of an input pulse.

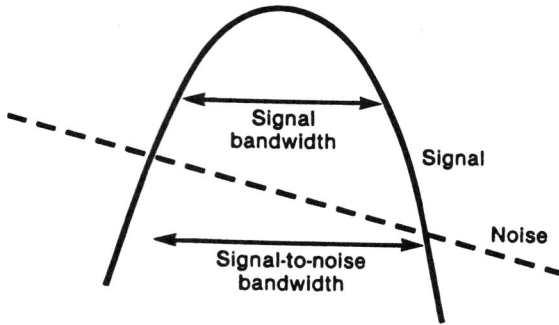

Fig. 11-5 Important distinction between signal bandwidth and signal-to-noise bandwidth.

noise. Applied to the data, it does as expected (Fig. 11-6d); the combination now has a flat spectrum. True, the signal spectrum is flat over the signal-to-noise bandwidth from f_1 to f_2; in this sense the signal bandwidth has been extended somewhat. But the noise has been *amplified* at the low and high frequencies; the overall effect of the deconvolution is harmful.

The standard solution to this situation is to follow the deconvolution with a band-pass filter (Fig. 11-6e), with its cutoff frequencies at f_1 and f_2. Clearly, there is no point in maintaining the processing response at frequencies where the input is predominantly noise. We see, then, the limitation of deconvolution in improving the vertical resolution:

> **Deconvolution allows us to extend the signal bandwidth out to the signal-to-noise bandwidth, but no further.**

The value and the limitations of deconvolution are seen clearly in Fig. 11-7. At (a) we see the signal just poking its nose above the noise, and at (b) the result of deconvolution and filtering. Both figures correspond to Fig. 11-6. In Fig. 11-7c there is more signal; perhaps we have just increased the fold of stack. The signal spectrum in Figs. 11-7a and c, and hence the signal bandwidth, is *not changed*. Figure 11-7c, however, offers a greater signal-to-noise bandwidth; after deconvolution and filtering, we have Fig. 11-7d. The signal bandwidth at (d) is much better than that at (b).

> **The function of the field work is to provide good signal-to-noise bandwidth at the reflection time where good vertical resolution is required. The important thing in the field is signal-to-noise bandwidth, not signal bandwidth. Given good signal-to-noise bandwidth, deconvolution can do the rest, and give us good signal bandwidth.**

This understanding of the different contributions of the field work and the processing is extremely important.

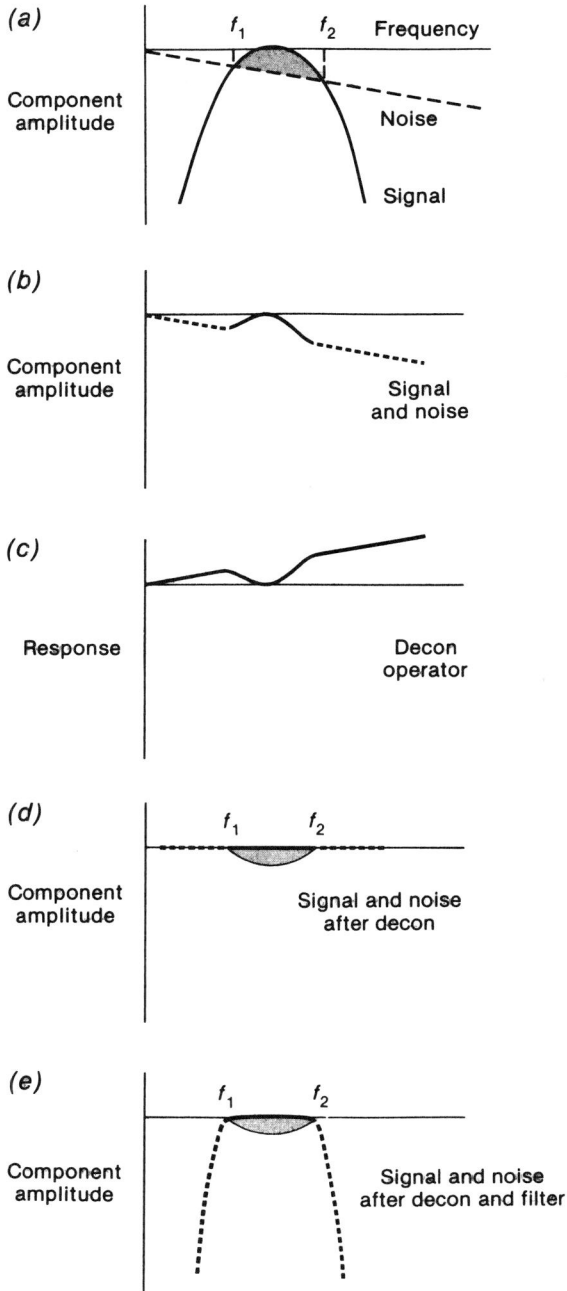

Fig. 11-6   Action of statistical deconvolution on a signal, in a background of wide-band noise.

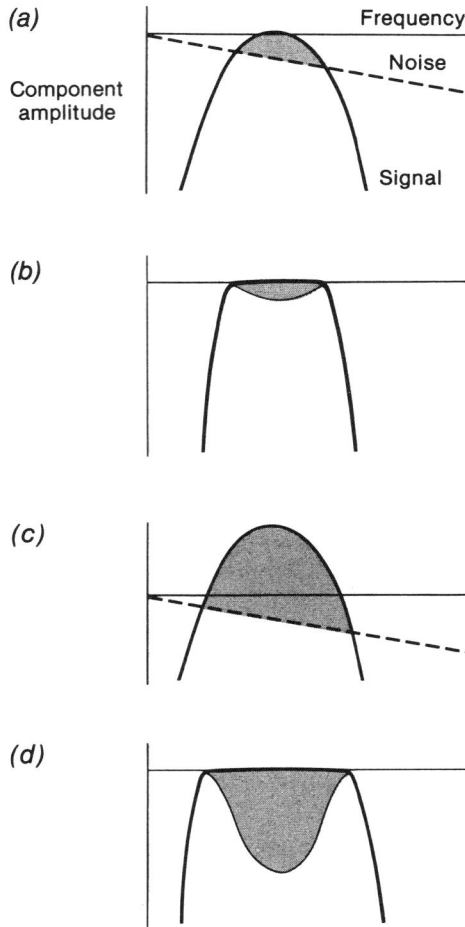

Fig. 11-7 How good signal-to-noise bandwidth affects deconvolution. (a) The signal and background noise of Fig. 11-6a. (b) The deconvolution output of Fig. 11-6e. (c) The signal bandwidth of (a), this time accompanied by good signal-to-noise bandwidth. (d) The deconvolution output based on the input of (c).

11.2 THE LINEAR SWEEP

Let us now see how we set about improving the resolution, while maintaining the *linear* sweep of all the previous discussion. We recall from Fig. 4-1a (reproduced as Fig. 11-8a) that the linear sweep has a constant rate-of-change of frequency; also, given constant amplitude (b), it has a flat amplitude spectrum (c). Since the spectrum of the auto-correlation function is the square of the spectrum of the sweep, the spectrum of the auto-correlation function is also flat.

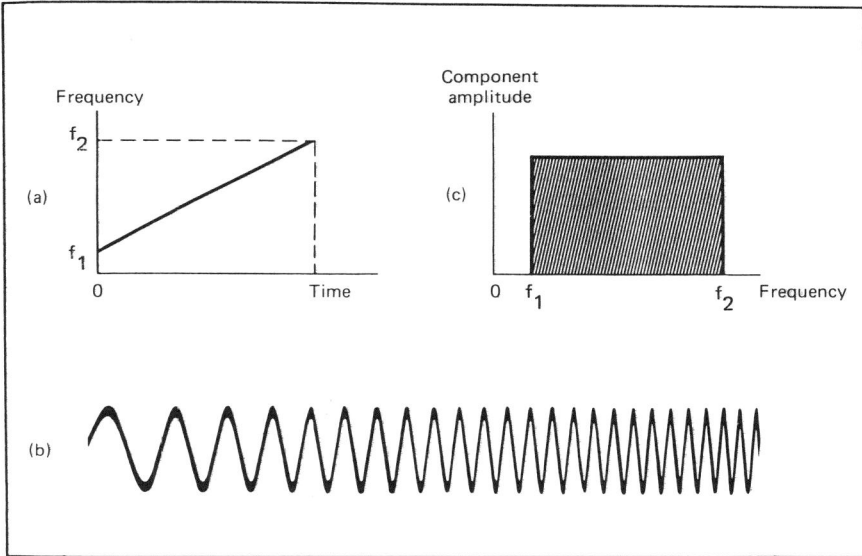

Fig. 11-8   Reminder (from Fig. 4-1) of the characteristics of the linear sweep.

Provided that the sweep is broad enough, therefore, we effectively have the situation of Fig. 11-4.

Our first move, then, is to increase the bandwidth of the linear sweep. Let us suppose that we have previously done a reconnaissance survey in the area and identified a target; now we wish to resolve the details of that target.

In Fig. 11-9a we see the probable situation realized by the reconnaissance survey. The target reflection was just far enough above the noise (after mute and stack) to have a signal-to-noise bandwidth equal to the full bandwidth of the sweep, 14 to 48 Hz. After deconvolution, the target reflection was as well resolved as it could possibly be with the bandwidth of 14 to 48 Hz.

The reconnaissance survey used eight sweeps, each 10 s long, from three vibrators. For the detail survey we seek better resolution at the target level, and decide to sweep double the bandwidth, 7 to 75 Hz. What happens if we still use eight sweeps, each 10 s long, from three vibrators?

We are spending only half the time in the original bandwidth of 14 to 48 Hz. The energy in this band is therefore less, and if the noise has not changed, the signal is now below the noise (Fig. 11-9b). The detail survey is worse than the reconnaissance survey.

To make the detail survey equal to the reconnaissance survey at target level, we must sweep for twice as long. Then the energy in the original band (14 to 48 Hz) is the same as before (Fig. 11-9c). However, *we have not yet done any good at target level;* after deconvolution and final filtering, the reflection bandwidth and the vertical resolution are the same as before.

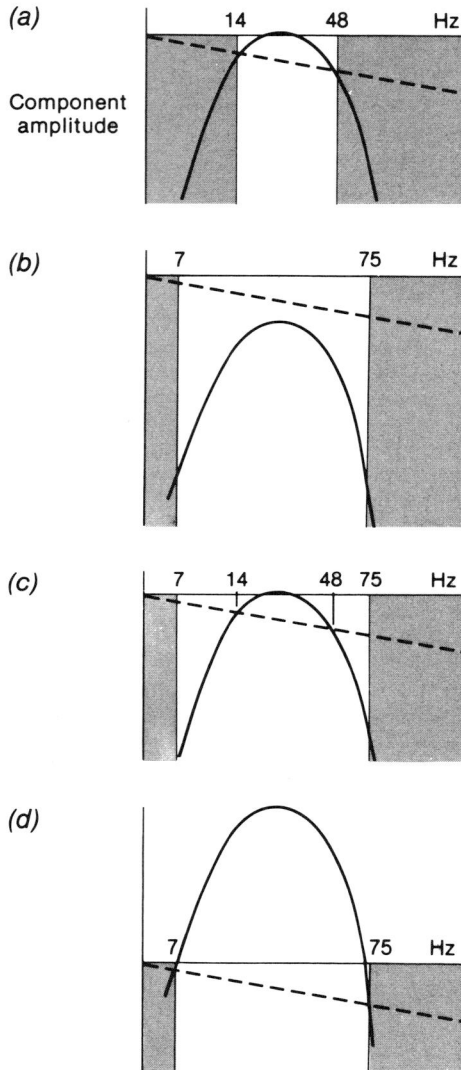

Fig. 11-9 Importance of sufficient signal-to-noise bandwidth. (a) Original sweep bandwidth and length. (b) Increased bandwidth, but the same sweep length. The signal spectrum falls below the noise because less time is spent at each frequency. (c) Increased bandwidth, but twice the sweep length. This brings the original bandwidth back to its original level. (d) Desired signal-to-noise bandwidth, achieved by sweeping much more than twice as long as the original.

It is true that the resolution at *higher* levels may well be better. There we expect a stronger signal and a better signal-to-noise bandwidth. With the original sweep of 14 to 48 Hz, no benefit could be derived from this, because the signal bandwidth was totally limited to the sweep frequencies. Now, however, with a wider sweep, we can expect the recoverable bandwidth to improve at shallower depths, until on the very shallow reflections it is again limited by the sweep.

Returning to the target level, we can see (Fig. 11-9d) that to realize the full bandwidth of 7 to 75 Hz we must sweep *much more than twice as long*. We must sweep and sweep and sweep and sweep, to bring the edges of the signal bandwidth above the noise. Production will be severely curtailed, and the cost per kilometer will rise steeply.

One conclusion is very clear: before we try to improve the bandwidth by this costly approach, we must do everything we can by less expensive methods.

Referring again to the agencies that limit the bandwidth (Fig. 11-10, a reprise of Fig. 11-2), we see that the first move is to use the stack-array method

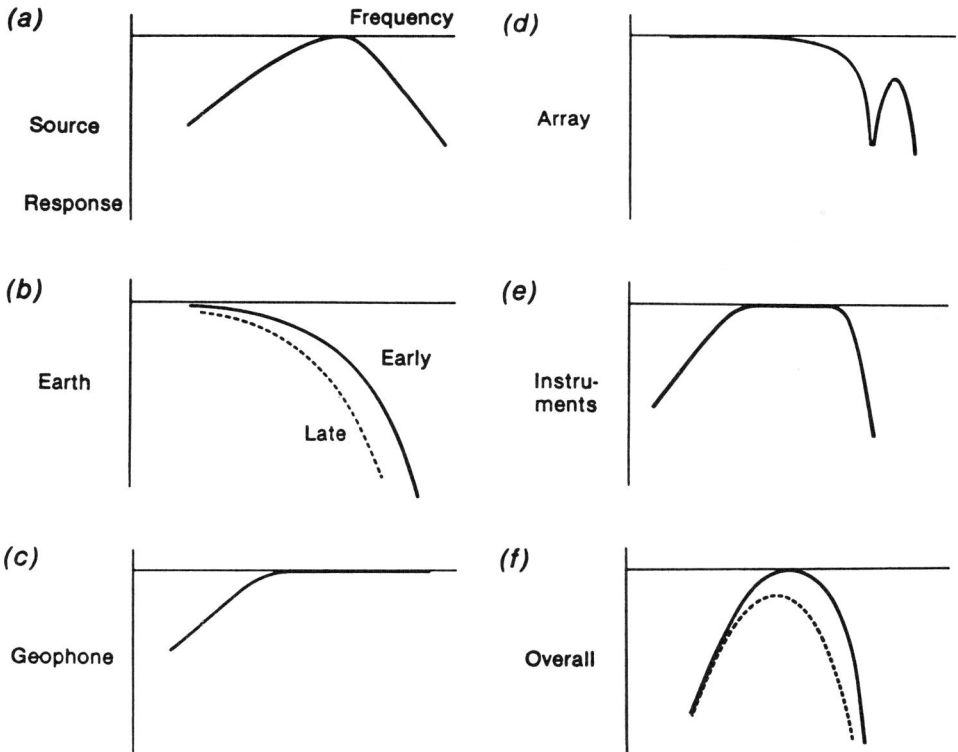

Fig. 11-10 Reminder (from Fig. 11-2) of the agencies that limit the reflection bandwidth.

of suppressing the source-generated noise, rather than long arrays of traditional type. This removes the array problem of Fig. 11-10d. The stack-array approach means, we remember, a split spread, a group length usually equal to the group interval, a source interval equal to the group interval, and source points halfway between group centers. In the sense that the highest frequency of the sweep is now the highest maintainable frequency through the recording and processing (the f_m of Chapter 10), this frequency now enters the determination of the group interval, as $V/2f_m \sin \theta$. As such, it defines the whole recording geometry.

Given sufficient attenuation of the source-generated noise in the stack array, we might next be led to lower the cutoff frequency of the geophone and the instruments (Fig. 11-10c and e). However, this is not necessary. The low-cut action, while it changes the signal bandwidth, *does not change the signal-to-noise bandwidth;* therefore, this low cut is immaterial to the final signal bandwidth after deconvolution.

Figure 11-11a takes our basic picture of signal and noise, and adds some typical ground roll, of high amplitude and low frequency. In Fig. 11-11b we see the geophone response we would like to use in this situation: a fairly high cutoff frequency, yielding a marked attenuation of the ground-roll frequencies (c). Of course, this response also cuts the low signal frequencies and the low noise frequencies. However, it does not change the *ratio* of signal to ground roll, or of

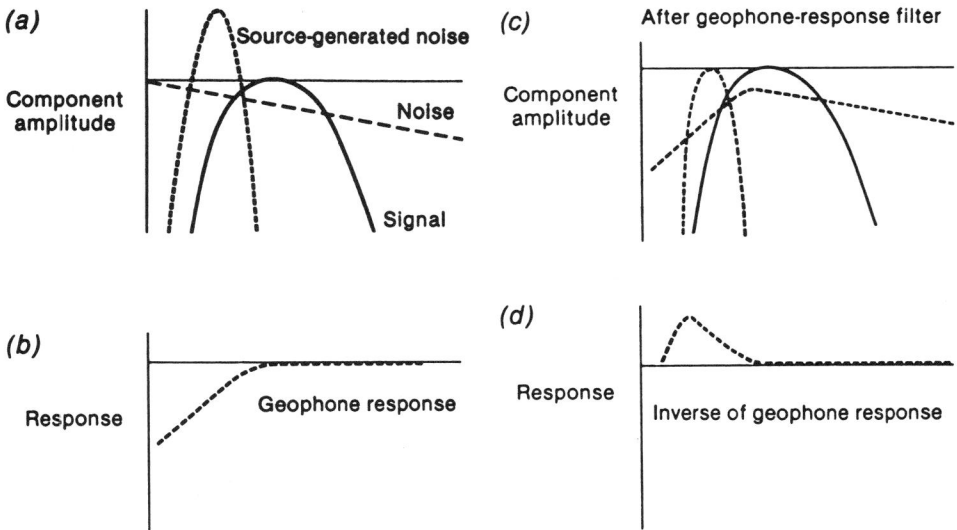

Fig. 11-11 Low-frequency cut of the geophone response. (a) Spectra of the signal, the ambient noise, and the source-generated noise. (b) Geophone response. (c) Effect of the geophone response on the spectra of (a). (d) Required inverse to the geophone response, modified for stability.

signal to other geophone noise. Therefore, in processing we can apply an inverse filter for the geophone response (d), and recover the original signal, the original noise, and the original ground roll (a).

In practice, of course, we must limit the inverse response of the geophone and roll it over (d); otherwise, it would have infinite amplification at zero frequency. We choose to do this at the frequency for which the signal falls to the *system* noise in the amplifier and digitizer; in a modern system, this certainly allows us to compensate for at least 20 dB of low-frequency attenuation in the geophone. Therefore, we can place the cutoff frequency of the geophone well above the low-frequency limit of the signal bandwidth and suffer no harm. Indeed, this does considerable good; the geophone is more rugged, its distortion is less, the ground roll does not dominate the gain control of the amplifier, the signal is thus recorded with better precision, and the reflections are more evident on the monitor record.

Since the limiting slope of the geophone response is 12 dB/octave, a 20-dB boost in the inverse geophone filter allows us to use a geophone with a natural frequency of 14 Hz and to recover with confidence frequencies down to about 5 Hz.

Thus it is no longer true that the low-frequency content of the reflections is constrained by the natural frequency of the geophones and by the need to apply a low-cut filter to the source-generated noise. In the field, we are free to attenuate the source-generated noise temporarily by the geophone response (and even, if desired, by an instrumental low-cut filter) in order to pass it through the recording system optimally; the source-generated noise is then suppressed in the stack array during processing, and the low signal frequencies are restored by an inverse filter.

The next band-limiting agency we should consider is the high-cut filter in the instruments (Fig. 11-10e). Again this does not affect the signal-to-noise bandwidth, and so can be partially compensated for in processing.

However, it costs us no more than the price of tape to raise the cutoff frequency of the field filter, and so this is what we do.

After attention to these matters, we are left with the vibrator and its coupling to the ground (Fig. 11-10a) and the earth itself (b). There is not much that we can do about the earth with a linear sweep; here we shall just accept it as a problem, and go on to the vibrator and its coupling. The frequency response of the vibrator, as we remember from Chapter 6, is limited by the stroke, the pump, the servo-valve, and the compressibility of the oil. For a high-resolution survey, we certainly use vibrators of the best frequency response available; we also stress good maintenance of the servo-valves, so that this frequency response is preserved during the survey. The frequency response of the vibrator–ground coupling, we remember, can be made more uniform by ground-force control; in principle, this is clearly desirable for high-resolution work.

However, in thinking about the frequency responses of the vibrator and its coupling, we must not lose sight of absolute power output. A flat response is no advantage if it is achieved merely by reducing the drive at the middle frequencies. Ideally, we seek to flatten the final output by increasing the drive at the edge frequencies; if this would produce overheating of the vibrator, we are content to take some reduction at the middle frequencies in exchange.

Whatever the frequency response of the vibrator and the coupling, we never forget the importance of keeping all the vibrators working, and of keeping the overall drive levels as high as practicable. And, just as it is important to maximize the signal, so it is important to minimize the noise. On a high-resolution survey we are even more concerned than usual to ensure quiet geophone plants and a minimum of traffic noise.

In our enthusiasm for the sophisticated things we can do with Vibroseis, we do not forget the simple and "obvious" factors; we keep the drive level up and the noise level down.

Also, in connection with the frequency response of the system, we see again the folly of applying amplitude tapers to the ends of the sweep. The tapers should be as short as possible, compatible with the avoidance of transient "bumps" in the vibrator output. It is also pointless and harmful to apply tapers to the sweep used in correlation. The earth supplies more than enough tapering (that is, filtering). Even if it did not, we would apply the filtering after deconvolution, not in the field or in correlation.

In summary, then, the high-resolution Vibroseis survey with a linear sweep seeks to minimize the loss of bandwidth in the arrays, the vibrator, and the vibrator–ground coupling. It also seeks to maximize the vibrator effort and to minimize the noise. When those things are done (and only then), it sets about building up the reflection signal—with more and more sweeps—until it achieves a signal-to-noise bandwidth equal to the sweep bandwidth. We might call this a *brute-force approach* to obtaining improved vertical resolution; it accepts that the final reflection signal will be peaked by the earth (as in Fig. 11-12a) and

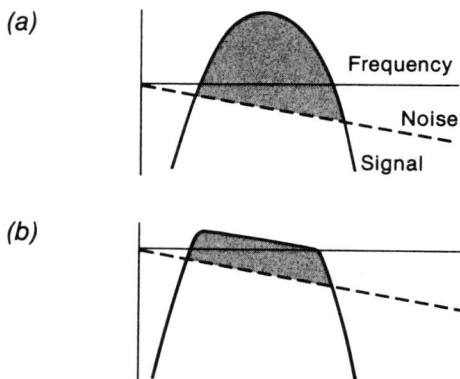

Fig. 11-12 We do not really need the signal to be head and shoulders above the noise; just the shoulders would be sufficient.

accepts that a great deal of sweeping time will be necessary to raise the edge frequencies of the reflected sweep above the noise level.

Obviously, the thing that is wrong with this brute-force approach is that much of the time spent sweeping the middle frequencies is wasted. At these frequencies the reflected signal is far above the noise—unnecessarily far. The energy used would be less and the deconvolution would be easier if the spectrum of the reflection signal were flatter (like Fig. 11-12b), while still maintaining signal-to-noise bandwidth over the full range of the sweep. This is the consideration that leads us to the nonlinear sweep.

11.3 THE NONLINEAR SWEEP

The elegance of Vibroseis rises to even greater heights with the nonlinear sweep. The total ability of Vibroseis to control the frequencies emitted by the source, and their relative strengths, extends dramatically our chances of improving the vertical resolution. In principle, as we shall see, it offers us whatever resolution we are prepared to pay for.

The basic concept of the nonlinear sweep is that the vibrator sweeps slowly through the frequencies we need to strengthen and quickly through those whose strength is already sufficient. The sweep rate is no longer constant.

Figure 11-13a compares the frequency–time relation for a linear sweep and for a specimen nonlinear sweep. (In this figure we depart from the usual practice of showing time along the abscissa, since in the practice of nonlinear sweeps it is more useful to regard frequency as the independent variable.) Both sweeps are 10 to 100 Hz, and both are 20 s long. The left-hand time axis represents an upsweep, the right-hand axis a downsweep. The nonlinear sweep is said to "dwell high," since it spends most of its time at the higher frequencies; we can see that it reaches 70 Hz in about 10 s, and then spends the remaining 10 s sweeping from 70 to 100 Hz.

The sweep rate is shown at (b). At 10 Hz it is about 25 Hz/s and at 100 Hz about 2.5 Hz/s. The sweep rate of the linear sweep, in contrast, is constant at 4.5 Hz/s.

And the spectrum? We remember that we have two interests: the spectrum of the sweep itself, and the spectrum of the effective Vibroseis "shot," which is the auto-correlation function of the sweep. We also remember that the spectrum of the auto-correlation function is the square of the spectrum of the sweep.

With that established, the answer could scarcely be easier. *The spectral amplitude of the auto-correlation function is inversely proportional to the sweep rate.* The spectrum of the sweep itself is the square root of this.

Not only is this simple, but it is intuitively very satisfying. It says, in effect, that *the frequency content of the Vibroseis "shot" is directly proportional to the time that the sweep spends sweeping through 1 Hz.*

(a)

Sweep

(b)

Sweep rate

(c)

Spectrum of auto-correlation
function of sweep

(d)

Spectrum relative to
linear sweep

(e)

Spectrum on logarithmic
frequency scale

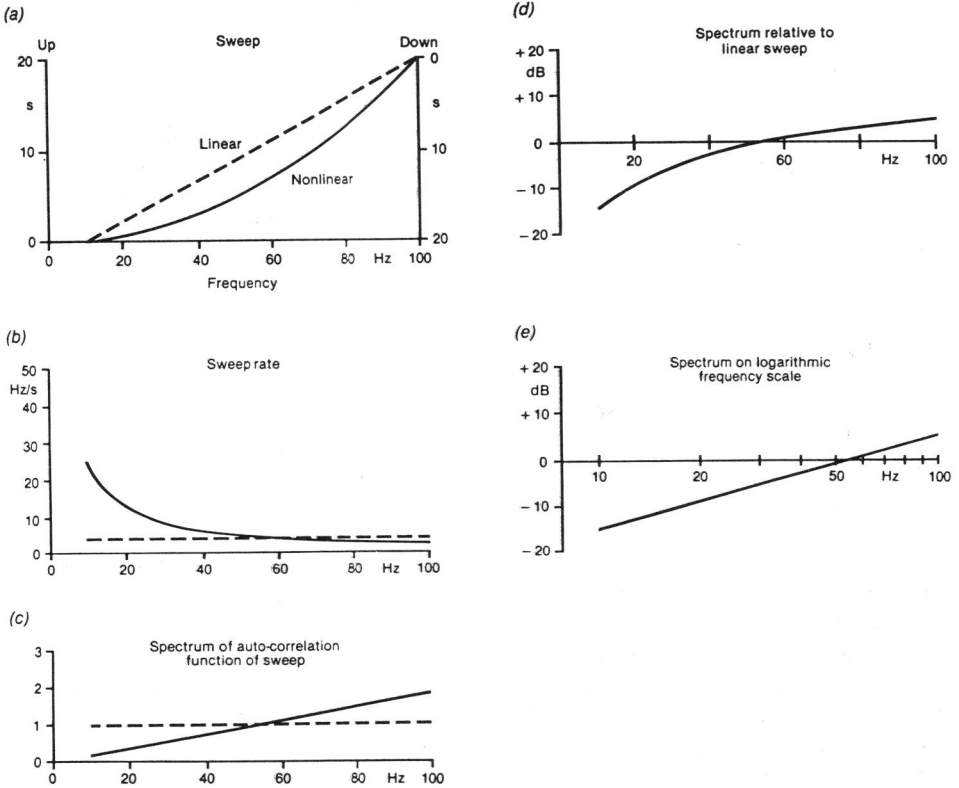

Fig. 11-13 Example of a nonlinear sweep. (a) Relation between sweep frequency and time. (b) Sweep rate as a function of frequency. (c) Spectrum of the auto-correlation function of the sweep. (d) Spectrum, relative to the linear sweep, plotted on a decibel scale. (e) The same on a logarithmic frequency scale. (In this and the following figures, ripples on the spectrum at the limits of the sweep are ignored, as they are not important for present purposes.)

In Fig. 11-13c, we take the reciprocal of the sweep rate of (b) and plot it as the spectrum of this effective Vibroseis signal. Obviously, our nonlinear sweep has been designed to yield a signal whose spectrum rises in direct proportion to frequency (that is, at +6 dB/octave). Also shown, and used as the reference, is the rectangular spectrum of the linear sweep. The figure illustrates once more the important principle that there is no free lunch; the slow sweeping through the high frequencies does indeed enhance the high frequencies, but at the expense of the lows. The gained area of the nonlinear spectrum above the linear spectrum at high frequencies is exactly counterbalanced by the lost area at low frequencies. The crossover point for this sweep is at the center frequency, 55 Hz.

Figure 11-13d plots the same information as (c), but on a decibel scale relative to the linear sweep. At (e) we change to a logarithmic scale of frequency; again we see that our Vibroseis "shot" has a spectrum rising at 6 dB/octave.

Let us suppose, then, that some part of our seismic system (the vibrators, for example, or the earth) has a predictable frequency response, which falls at 6 dB/octave. We can precompensate for this fall by using the nonlinear sweep of Fig. 11-13a. The signal goes out preemphasized in the high frequencies, and comes back flat.

Obviously, this is important, and we must develop it. In particular we must ask: Given a known or measured frequency response, in the equipment or in the earth, how can we design a nonlinear sweep to compensate for it? First, we must narrow the options, because there is an infinity of nonlinear sweeps. We find that for many purposes we can narrow to just two classes of nonlinear sweep: decibel-per-octave sweeps and decibel-per-hertz sweeps.

11.3.1 Decibel-per-Octave Sweeps

These are sweeps that yield an effective Vibroseis "shot" whose spectrum varies with a power of frequency. They are appropriate to the precompensation of any effect that can be expressed as so-many decibels per octave; we remember that $+6$ dB/octave represents a response proportional to frequency; $+12$ dB/octave, to the square of frequency; -6 dB/octave, to the inverse of frequency; and so on.

Appendix B sets out the equations governing such sweeps, and Fig. 11-14 illustrates four of them graphically. First (a), we see the one we have already met in Fig. 11-13; it is now labeled $+6$ dB/octave. To it we add the sweep representing $+12$ dB/octave, which represents a major boost of the high frequencies. Then we also add the corresponding boosts of the low frequencies, -6 and -12 dB/octave.

As before, we keep the linear sweep (dashed) as a reference; again we see that there is no free lunch. The stronger we make the frequencies at one end of the spectrum, the weaker become those at the other. And if we ask too much at one end, the sweep rate at the other becomes so great that the vibrator electronics have difficulty following it. In Fig. 11-14b we see that, below about 18 Hz, the $+12$-dB/octave sweep rate exceeds 50 Hz/s; this may be too much for the phase-lock circuitry.

We also see yet another warning against applying excessive tapers in the field. At (a), we see that a $+12$-dB/octave sweep with a 1-s taper would not reach full amplitude until about 47 Hz. Yes, 47 Hz. Even with a 0.1-s taper, full amplitude is not reached until 17 Hz, and much of the bottom half-octave of the sweep is thrown away. For nonlinear sweeps in particular, we learn once again that tapers should be as short as the vibrator allows, and that no taper should be applied to the sweep used in correlation.

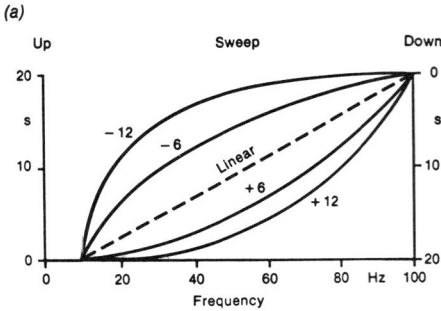

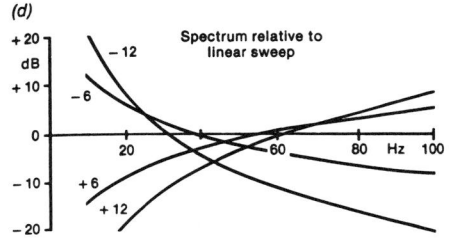

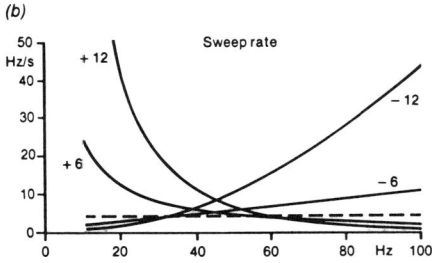

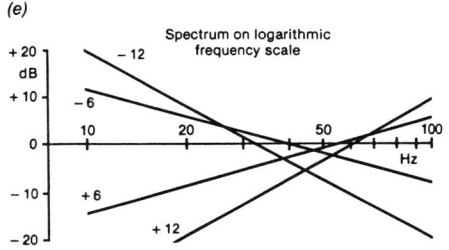

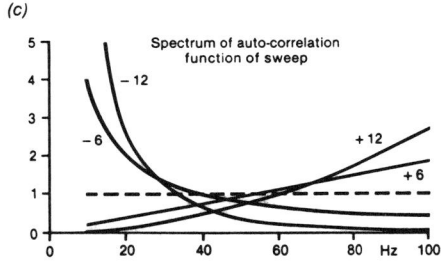

Fig. 11-14 Counterparts of Fig. 11-13 for nonlinear sweeps yielding auto-correlation functions whose spectra rise at 6 and 12 dB/octave, and fall at −6 and −12 dB/octave. The energy in all the sweeps is the same as that of the linear sweep.

11.3.2 Decibel-per-Hertz Sweeps

These are sweeps that yield an effective Vibroseis "shot" whose spectrum varies by the same proportion across each 1 Hz of frequency. They are appropriate to the precompensation of any effect that can be expressed as so-many decibels per hertz. Important among these, as we shall see, is the effect of absorption in the earth.

Appendix C sets out the equations governing these sweeps, and Fig. 11-15 illustrates a representative one graphically. In fact, this sweep, with its rising

(a)

Up   Sweep   Down

(b)

Sweep rate

(c)

Spectrum of auto-correlation
function of sweep

(d)

Spectrum relative to
linear sweep

(e)

Spectrum on logarithmic
frequency scale

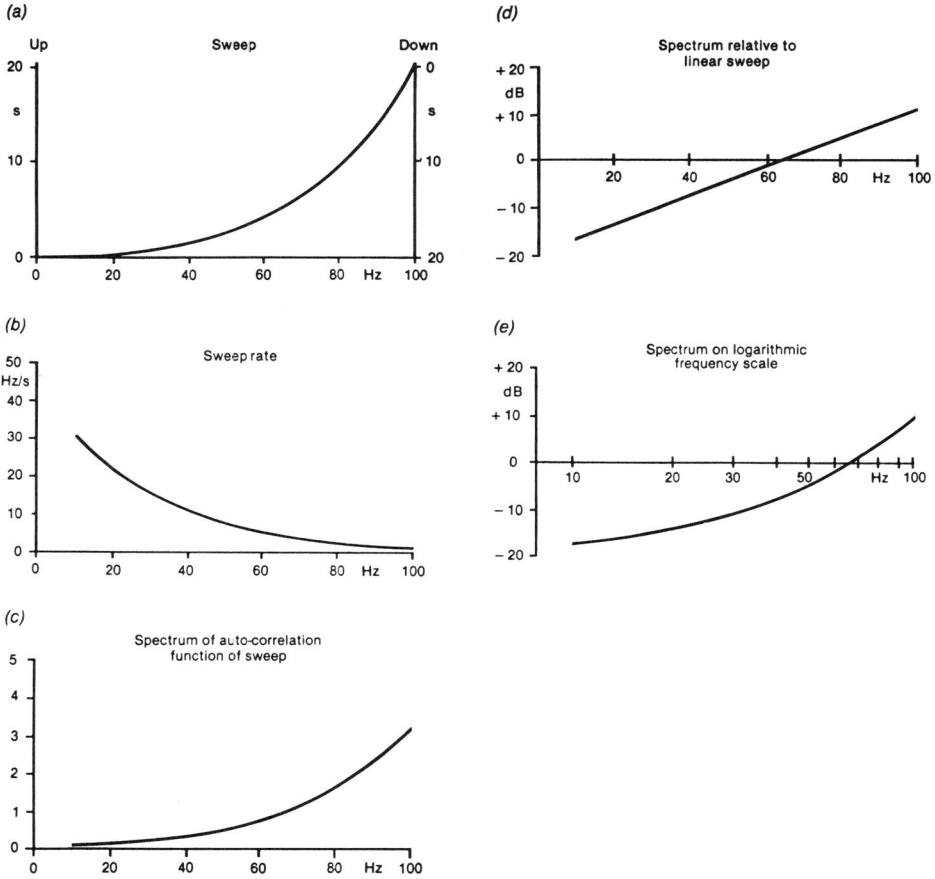

Fig. 11-15 Counterparts of Fig. 11-13 for a nonlinear sweep yielding an auto-correlation function whose spectrum rises at 6 dB per 20 Hz.

response, is of the type appropriate to the precompensation of absorption; we shall return to it later.

Figure 11-14e showed us that the precompensation given by a decibel-per-octave sweep is linear when plotted on a logarithmic frequency scale; Fig. 11-15d shows us that the precompensation given by a decibel-per-hertz sweep is linear on a linear frequency scale. Whereas the +6-dB/octave curve of Fig. 11-14 gave a high-frequency boost of 6 dB for every octave of frequency, the sweep of Fig. 11-15 gives a high-frequency boost of 6 dB for every 20 Hz of frequency.

We have to choose. Let us decide what to do by considering some examples.

11.3.3 Compensating the Vibrator

Figure 11-16a reminds us that the low-frequency output of a vibrator is constrained by the need to prevent the piston from hitting the stops. Let us say that the output falls at 6 dB/octave below 20 Hz, and that we wish to compensate this down to 10 Hz. The high-frequency output is constrained first by the fact that the vibrator is a constant-force device; let us say that the output falls at 6 dB/octave above 50 Hz, and that we wish to compensate for this up to 100 Hz. Then the problem is to design a nonlinear sweep whose auto-correlation function has the spectrum of Fig. 11-16b.

Again, we use as the reference a linear sweep from 10 to 100 Hz over 20 s. For the frequencies from 20 to 50 Hz, where no compensation is required, we maintain this linear sweep; the sweep rate is constant at $(100 - 10)/20$, or 4.5 Hz/s.

Now let us look at the $+6$-dB/octave boost required at the high frequencies. Using the equations in Appendix B (for a low frequency of 50 Hz, a high

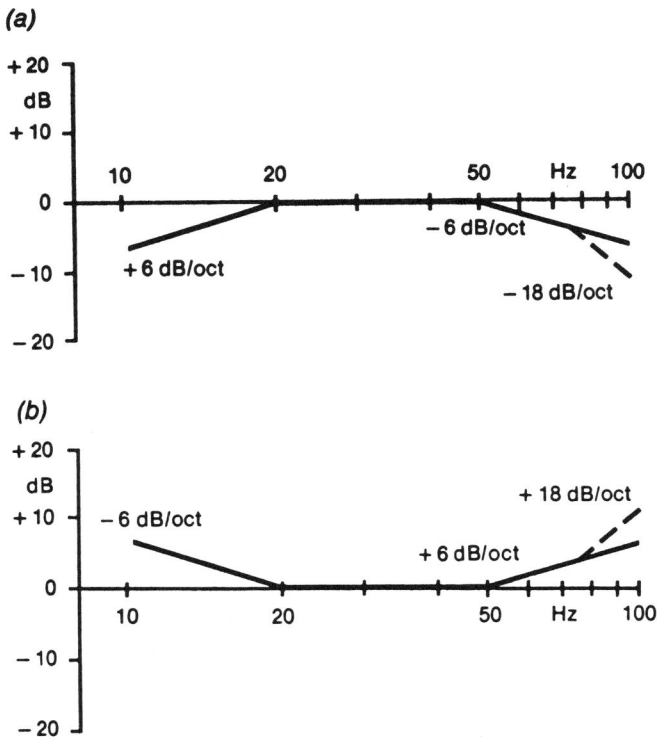

Fig. 11-16 Compensating the vibrator response. (a) Likely response for the vibrator itself. (b) Characteristic required of a nonlinear sweep to correct this response.

frequency of 100 Hz, and a limiting sweep rate of 4.5 Hz/s at 50 Hz), we can quickly compute that the sweep we need above 50 Hz is

$$f = (450t + 2500)^{1/2}$$

Similarly, at the low-frequency end, we enter the equations with a low frequency of 10 Hz, a high frequency of 20 Hz, and a limiting sweep rate of 4.5 Hz/s at 20 Hz. We compute[1] that the sweep we need below 20 Hz is

$$f^{0.01} = 0.0023t + 1.023$$

Figure 11-17a shows the final sweep. Also shown, dashed, is the reference linear sweep between the same end frequencies. Not only do we see graphically that there is no free lunch, but we can now compute the actual price. The linear sweep takes 20 s to sweep from 10 to 100 Hz; the nonlinear sweep takes 26.4 s.

We can also see that the high-frequency compensation is much more expensive than the low-frequency compensation (although both are 6 dB/octave over one octave). In fact, the extra time needed for the low-frequency compensation is less than 1 s, whereas at the high-frequency end it is about 5 s. We shall return to this point again: *low-frequency compensation is both easy and cheap.*

In Fig. 11-16a we included not only the two 6-dB/octave constraints on the vibrator output, but also (dashed) the 18-dB/octave high-frequency loss that arises when we consider the elasticity of the oil. In general-purpose vibrators, this starts to take effect at about 80 Hz (as shown), although in vibrators configured for high-resolution work the frequency is likely to be higher. What happens if we try to compensate for this?

Figure 11-17b gives the answer. The total length of the sweep is now nearly 29 s. Of this, more than one-third is spent between 80 and 100 Hz.

Now let us consider a quite different application of the nonlinear sweep to the compensation of the vibrator. We remember from Chapter 8 that at very low frequencies the output of the vibrator is corrupted by harmonic distortion. This may become extreme if the baseplate actually chatters (or *decouples*) on a hard surface. One solution, we remember, is to turn down the drive, automatically, at the frequencies where decoupling occurs. This overcomes the problem, but leaves the reflections deficient at the low frequencies. To restore these frequencies, we may design an appropriate nonlinear sweep.

Let us suppose that Fig. 11-18a represents a constant-amplitude linear sweep; we take it as a downsweep from 100 to 10 Hz. We find that to stop baseplate decoupling we must apply the amplitude reduction of Fig. 11-18b: from an amplitude of 1 at 30 Hz down to an amplitude of 1/4 at 10 Hz. Some sweep generators can do this reduction linearly with *time only* (not frequency). Then the problem is to compensate the diminished amplitude at low frequencies by a diminished sweep rate, and so to restore a flat spectrum.

[1]Using an *n* value of −0.99.

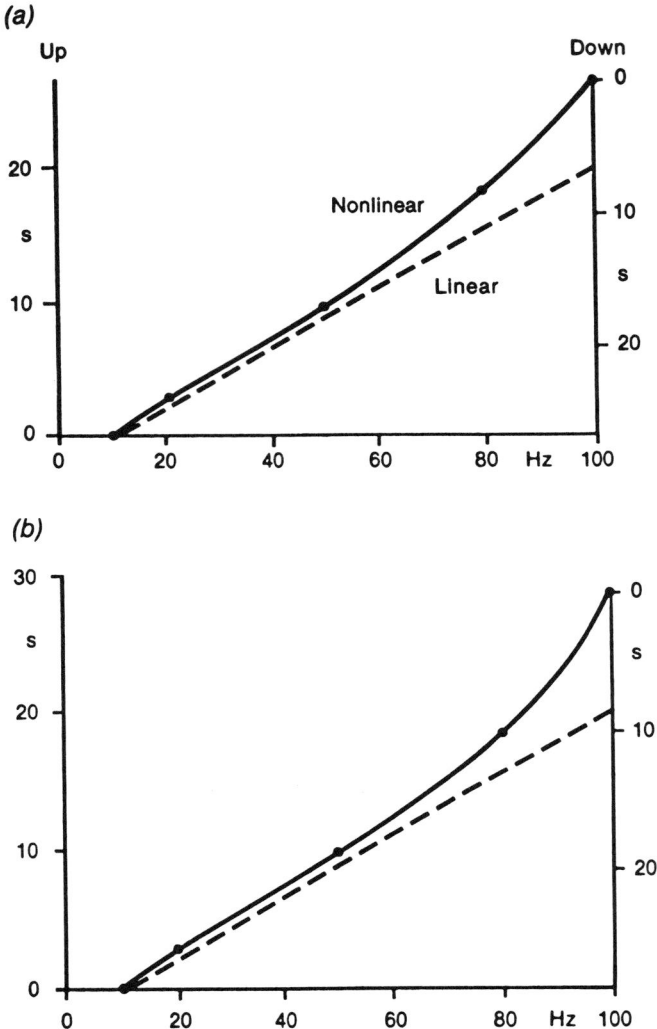

Fig. 11-17 Sweeps required to compensate the vibrator responses of Fig. 11-16a. (a) The nonlinear sweep that assumes the high-frequency droop of 6 dB/octave. (b) The nonlinear sweep that assumes the high-frequency droop of 6 dB/octave from 50 to 80 Hz and 18 dB/octave above 80 Hz.

First, we must note that, although we apply the low-frequency amplitude reduction to the sweep emitted by the vibrators, it would be foolish to do the same to the control sweep used in correlation. Since the auto-correlation process squares the amplitude spectrum of the sweep, it would be squaring the amplitude reduction to no benefit. Thus, the vibrator sweep of Fig. 11-18b, correlated with the control sweep of (a), yields the same correlation pulse as we

(a)

Amplitude

1 100 Hz   10 Hz

0

Time

(b)

1 100 Hz   30 Hz

10 Hz
0   ¼

(c)

1 100 Hz   30 Hz

10 Hz
0   ½

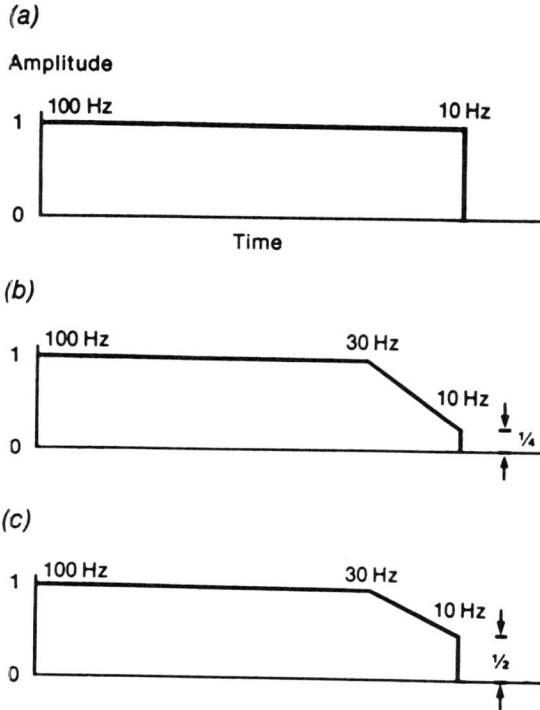

Fig. 11-18 One way to stop baseplate decoupling. (a) Control sweep, which maintains constant amplitude. (b) Sweep that prevents decoupling. The sweep emitted by the vibrators is reduced in amplitude to one-quarter at the lowest frequency. (c) The resulting correlated pulse is the same as if both the emitted and control sweeps were reduced to half-amplitude at the lowest frequency.

would have obtained had we used the sweep of (c) for both the vibrators and the control. The amplitude reduction is now the square root—to one-half instead of one-quarter. For calculation purposes, therefore, we say that a reduction to one-quarter on the vibrator sweep alone is equivalent to a reduction to one-half on the spectrum of the auto-correlation function.

The sweep we need is the decibel-per-hertz sweep of Fig. 11-15 and Appendix C, but now for a low-frequency boost. Figure 11-19c shows the desired inverse spectrum, and (b) the variation of sweep rate that achieves it. Figure 11-19a gives the sweep itself, as a relation between frequency and time; it is nonlinear up to 30 Hz and linear above 30 Hz.

If now the amplitude–time relation of Fig. 11-18c is applied to the downsweep of Fig. 11-19a, and the correlation is performed against the same sweep with constant amplitude, the resulting correlation pulse is the same as that of a constant-amplitude linear sweep. This has been achieved without the distortion and loss of signal associated with baseplate decoupling.

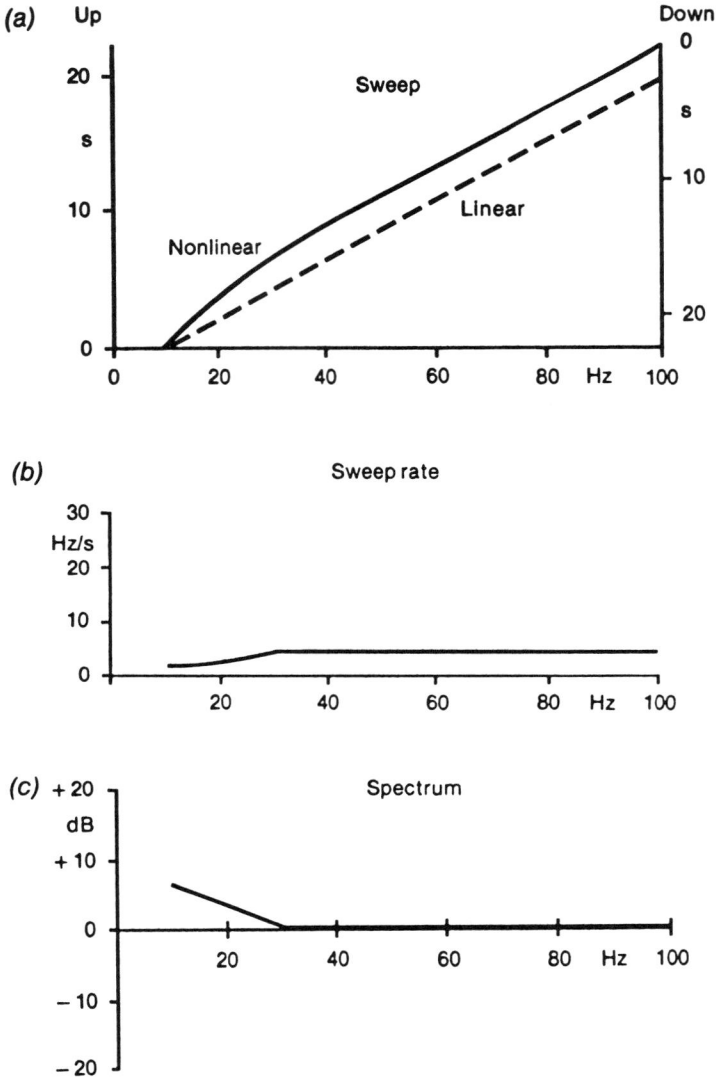

Fig. 11-19 Sweep designed to compensate a linear reduction of sweep amplitude below 30 Hz.

The price we pay is quite reasonable: the sweep duration has increased from 20 s to nearly 22 s. As we found before, the compensation of the low-frequency attenuation is easy and inexpensive. On any ground surface for which baseplate decoupling is a pervasive problem, therefore, a good solution is a reduction of amplitude at low frequencies and the compensation of this reduction by the appropriate nonlinear sweep.

We should also consider compensation of the vibrator–ground coupling. Figure 11-20 (a reprise of Fig. 6-5) reminds us of the effect. If we are prepared to accept one of these curves as representative of our prospect, we can approximate it by slopes of so-many decibels per octave and compute the corresponding nonlinear sweep.

We remember from Chapter 8 that modern vibrator electronics do their best to lock both the phase and the amplitude of the ground force to the control sweep. On the types of terrain for which the amplitude control works well, this reduces the need to use nonlinear sweeps for the compensation of the vibrator and, to some extent, of the coupling. Perhaps the future will bring vibrator electronics that measure the ground force and the induced motion, do what is practical to do electronically to keep the effective output constant with frequency, and then automatically introduce a calculated nonlinearity into the sweep to compensate the remaining departures from the ideal response.

11.3.4 Compensating for the Earth

We remember that there are three agencies in the earth that attenuate the high frequencies and one that attenuates the low frequencies.

The three that attenuate the high frequencies are absorption, short-path multiples, and scattering. The only one of these that is smoothly progressive and predictable is absorption; thus, in the design of nonlinear sweeps to compensate the attenuation, we usually concentrate on absorption.

We recall (Appendix C) that the absorption is specified in terms of so-many decibels per cycle, or equivalently in terms of Q (where 1 dB/cycle = $27/Q$). In the deep earth the absorption may be as low as 0.03 dB/cycle (Q about 1000), while in shallow unconsolidated materials it may exceed 1 dB/cycle (Q less than 30). The absorption associated with a particular path into the deep earth and back is thus the product of an average absorption value (perhaps 0.1 to 0.2 dB/cycle) and the number of cycles. We also recall that the number of cycles is itself the product of frequency and reflection time, so an absorption of 0.1 dB/cycle and a reflection time of 3 s have the same combined effect as an

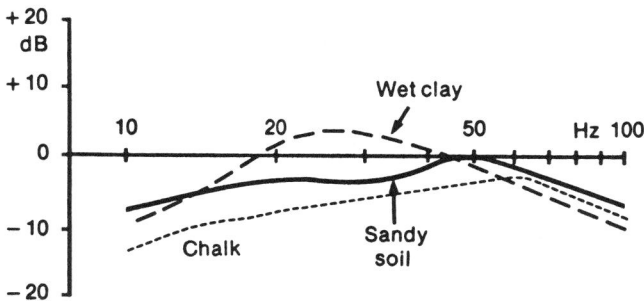

Fig. 11-20 Some typical responses for the vibrator–ground coupling.

absorption of 0.2 dB/cycle and a reflection time of 1.5 s; both situations represent 0.3 dB/Hz.

The sweep we need is the decibel-per-hertz sweep. In fact, Fig. 11-15 (repeated here as Fig. 11-21) is the decibel-per-hertz sweep appropriate to the above value of 0.3 dB/Hz. As we look at this, we must remember that the sweep depicted was designed to have the same end frequencies and the same duration as the reference linear sweep; in Figs. 11-21d and (e) the 0-dB value corresponds to the linear sweep.

It is important to note that, because absorptive attenuation is progressive with reflection time, the nonlinear sweep to compensate for absorption applies *only at target level*. Shallower reflections are overcompensated, and deeper reflections are undercompensated.

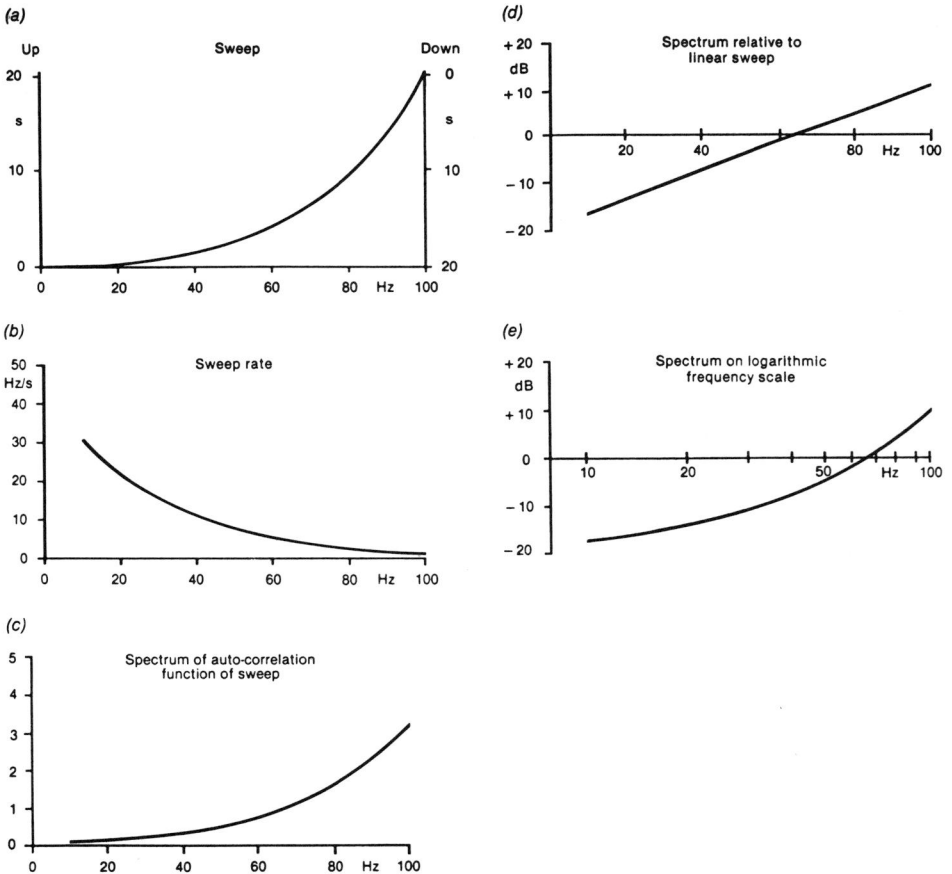

Fig. 11-21   Reminder of the action of the decibel-per-hertz sweep.

Now we turn to the agency that attenuates the low frequencies. This, we remember, is a consequence of the small size of the source relative to a wavelength. The baseplate itself is small at any reflection wavelength. Even with a source array of any normal dimensions, the source array is still small relative to a reflection wavelength over most of the frequency band. The result is that the earth, within a wavelength or so of the source, imposes a loss of 6 dB/ octave toward the low frequencies.

By now, this is a simple problem for us. We need a sweep designed to provide −6 dB/octave; this is the one labeled −6 in Fig. 11-14 (repeated here as Fig. 11-22). In contrast to the compensation of absorption, this compensation is appropriate to all reflections, whatever their travel time.

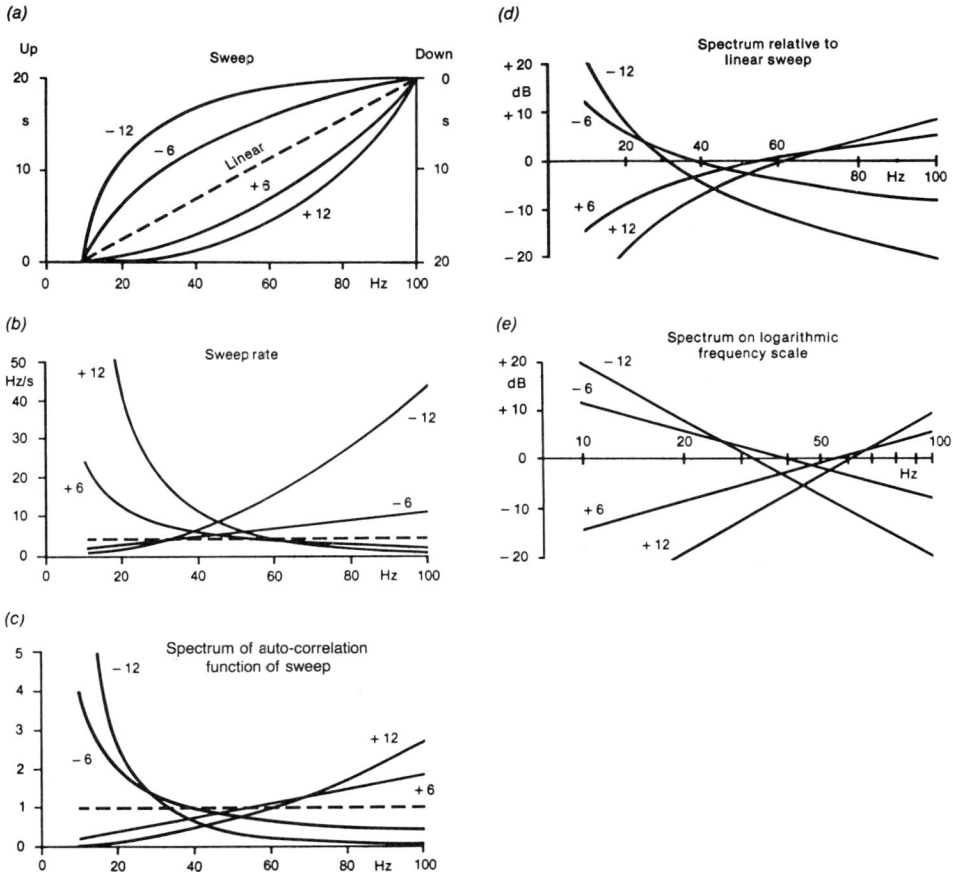

Fig. 11-22   Compensation of the low-frequency loss associated with a source that is small relative to a wavelength. This is done with the sweep labeled −6 dB/octave.

11.3.5 Compensating Both the Vibrator and the Earth

We now see the full potential of Vibroseis for improving the resolution. We can place Vibroseis on a par with dynamite by using a nonlinear sweep calculated to compensate the imperfect frequency response of the vibrator; in effect, we provide a short, sharp input pulse to the earth. Then we can begin to *improve* on dynamite, by using a nonlinear sweep calculated to compensate the frequency response of the earth.

Figure 11-23a brings together several of the resolution-limiting agencies we have discussed, and in (b) they are added to yield an overall response. The peak of this response is about 19 dB down, relative to the hypothetical case without earth effects; this loss, of course, is always present—for dynamite, linear sweeps, and nonlinear sweeps. Here we are concerned only with the nonlinearity of the sweep, so we set the peak in (b) to 0 dB, and deduce the corresponding relation between inverse sweep rate and frequency in (c).

In Fig. 11-23c we keep the sweep rate at the peak of the response equal to the 4.5 Hz/s that we used for the original linear sweep (10 to 100 Hz in 20 s). Given this variation of sweep rate with frequency, we can calculate the relation between sweep frequency and time, and the total duration of the sweep (Appendix D); the result is shown in Fig. 11-24. The nonlinear sweep is 53 s long; its linear counterpart was 20 s.

Going back to Fig. 11-23c and the right-hand scale, we can see that at 85 Hz the nonlinear sweep costs us 1 s of sweeping time for every extra hertz of compensated bandwidth. At 98 Hz it costs us 2 s for every hertz.

If we decide to use this nonlinear sweep (or indeed any variation on it), the computer can quickly calculate the sweep itself. The usual technique is to split the sweep duration into small time segments (typically 8 ms) and to calculate the sweep within each such segment as a linear sweep. Thus, to compute a downsweep corresponding to Figs. 11-23 and 11-24, the first segment is a linear downsweep starting at 100 Hz and having a sweep rate of 0.42 Hz/s. The initial phase angle is 0°. The calculated phase angle and frequency at the end of the first segment becomes the initial phase angle and frequency for the second segment, and that calculated frequency determines the sweep rate for the second segment.

When the sweep is thus calculated, it may be "burned" into several identical read-only memory chips; these chips are inserted into the sweep generators in the vibrators and the recording truck, and are thus used to control the outgoing signal.

If the sweep generators allow it, we may similarly preprogram the vibrator amplitude. Thus, we reduce the drive at low frequencies to eliminate baseplate decoupling, and modify the sweep to compensate this. Then we also increase the drive at high frequencies, to a limit imposed by the servo-valve or by overheating; this reduces the need for very small sweep rates at the high frequencies, and so reduces the loss of production.

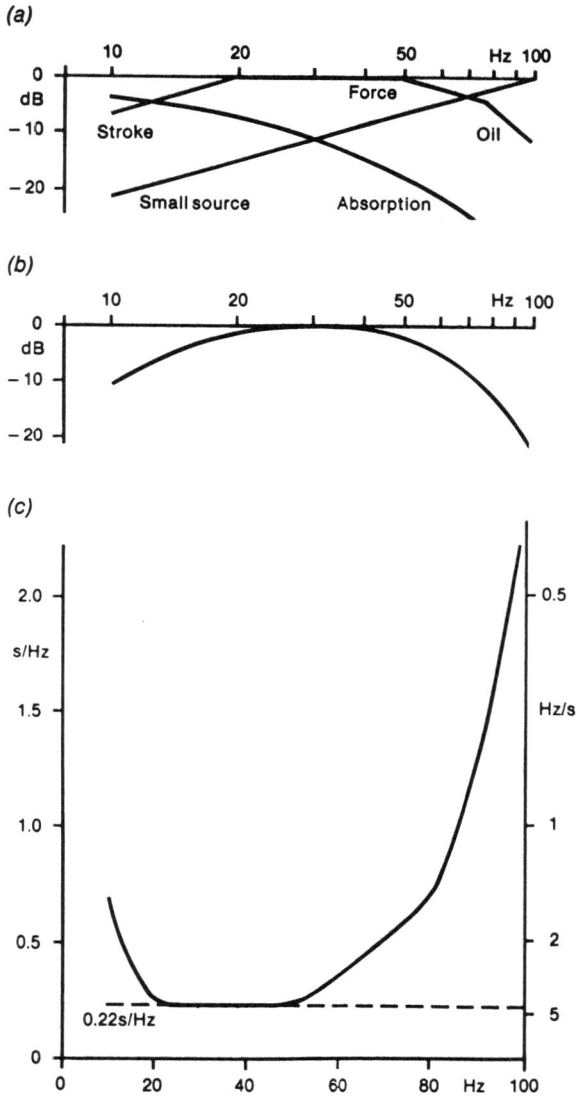

Fig. 11-23   Compensation for combined losses. (a) Some of the agencies that affect the spectrum of the reflected pulse. (b) Combined effect of the agencies in (a), normalized so that the peak of the spectrum is at 0 dB. (c) Variation of sweep rate with frequency for compensation of the combined factors.

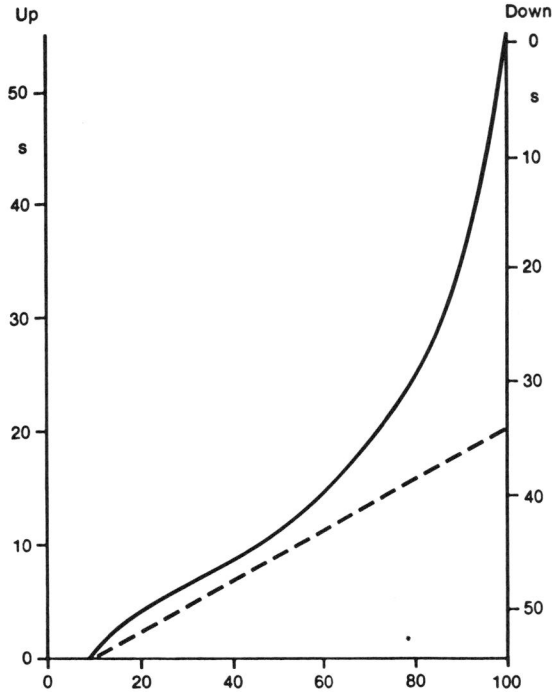

Fig. 11-24 Nonlinear sweep that compensates the attenuations of Figs. 11-23a and b.

At this stage, we are quite excited about nonlinear sweeps. The concept is simple and elegant, and the mathematics not too daunting. We can compensate the signal for any known loss, and all it takes is time. True, it takes quite a lot of time, particularly at the highest frequencies. But we have the choice of whatever signal bandwidth we are prepared to pay for.

Alas, the full truth is not quite so pleasant. We have forgotten a cardinal point that we learned at the beginning of this chapter: *the important thing in the field is not signal bandwidth, but signal-to-noise bandwidth.* When we ponder this, we find it imposes a major constraint on what we can expect from nonlinear sweeps.

11.3.6 Signal-to-Noise Considerations

Thus far, we have considered two limiting options for the design of nonlinear sweeps. In the first (exemplified by Figs. 11-13 to 11-15), we do not change the *sweep duration* when we go from a linear to a nonlinear sweep. We have seen clearly that this must mean a reduction in the sweep rate at the frequencies where no compensation is required; this in turn means that the effective output at these frequencies is weakened.

In the second (exemplified by Figs. 11-17 to 11-19, 11-23, and 11-24), we do not change the *sweep rate* at the frequencies where no compensation is required, and thus we maintain the effective output at these frequencies. We have seen clearly that this must mean the sweep is longer.

In the use of nonlinear sweeps, we face a choice: we can lose reflection signal at the middle frequencies, or we can lose production. There is no escape from this choice.

We make the choice, obviously, on considerations of signal-to-noise ratio. Let us go back to Fig. 11-9, modify it somewhat, and thus obtain Fig. 11-25. We recall that we started with a reconnaissance survey in Fig. 11-25a, achieving signal-to-noise bandwidth at the target level only over the band from 14 to 48 Hz. For the detail survey, we first changed to a sweep of 7 to 75 Hz; immediately we found that, even to maintain the original signal-to-noise bandwidth, we had to maintain the original sweep rate, and so double the sweeping time. This loss of production gained us nothing, however, because the signal in the regions of extended bandwidth was below the noise (Fig. 11-25b). To achieve a signal-to-noise bandwidth of 7 to 75 Hz with a linear sweep, we had to sweep and sweep and sweep, all day. When we had done it, we had unnecessarily high signal-to-noise ratios at the middle frequencies. This was grossly inefficient, and we longed for a way to realize the situation in (d).

The nonlinear sweep gave us that way. Preoccupied as we were with the *signal* spectrum, however, we elected to compensate the signal spectrum to flatness, and thus obtained Fig. 11-26a. The first point to emerge from a consideration of signal-to-noise ratios, therefore, is that if the noise does not have a flat spectrum the signal need not have a flat spectrum. This is all to the good, since the spectrum of seismic noise typically falls to the high frequencies.

But then—Oops!—we see that we have made a mistake. In Figs. 11-25d and 11-26a, *we have implicitly assumed that the nonlinear sweep has not affected the spectrum of the noise.* We have carried dynamite-type thinking into Vibroseis. (This is a fair way to express the fact, because that is exactly what the industry did in the early days of nonlinear sweeps.)

But the nonlinear sweep *does* affect the spectrum of the noise. We prepared ourselves for this conclusion in Chapter 10, where we noted that the amplitude of the correlated Vibroseis pulse varies directly with the duration of the sweep, while the amplitude of the noise (within the bandwidth of the sweep) varies with the square root of the duration of the sweep. This, of course, is why standard Vibroseis practice yields the same signal-to-noise ratio with 10 sweeps of 10 s or 5 sweeps of 20 s. If we consider a sweep bandwidth of 1 Hz, then, we improve the signal-to-noise ratio with the square root of the time we spend sweeping it, which is the reciprocal of the sweep rate.

It is important that we are comfortable with this. The longer we sweep, the more the signal—of course. But a longer sweep also takes in more noise.

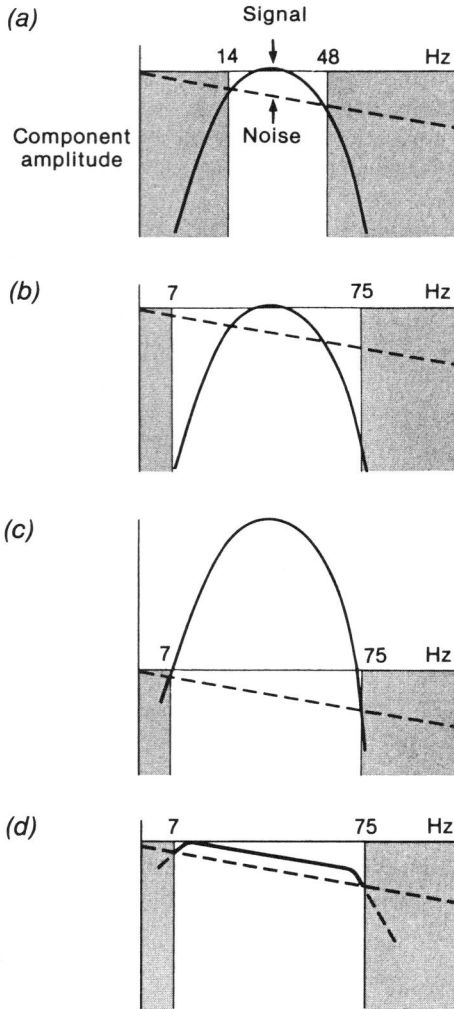

Fig. 11-25 Reminder that, for high-resolution work and a defined target level, the brute-force approach is less efficient than the shoulders-only approach.

At any frequency, the signal spectrum is defined by the time per hertz. The noise spectrum is defined by the square root of this. The signal-to-noise spectrum, therefore, is defined by the square root of the time per hertz.

Now we can see that the correct version of Fig. 11-26a is Fig. 11-26b. The nonlinear sweep, in applying to the signal a response that is inverse to the signal spectrum, has applied to the noise the square root of the same inverse response.

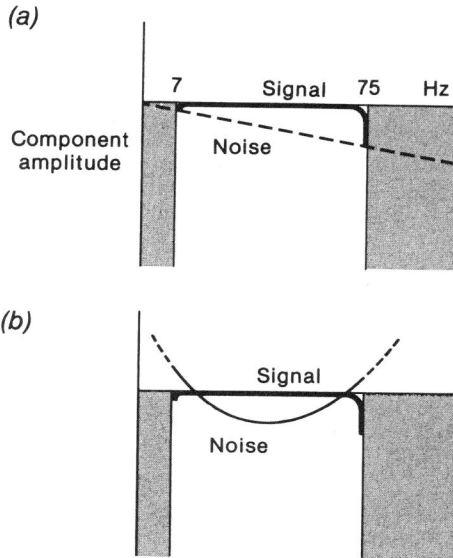

Fig. 11-26   Using a nonlinear sweep to flatten the signal spectrum. (a) Signal-to-noise spectrum of the sweep, as sometimes inferred in error. (b) Actual spectrum; the noise rises with the square root of the signal-flattening action.

The noise performance of the nonlinear sweep, therefore, is less than we hoped for. The low- and high-frequency extremes of the band, for which we fought so hard, have poor signal-to-noise ratios; they must be filtered out. This is the reason why many operators were disappointed by their first tests of nonlinear sweeps.

Figure 11-27 depicts the problem more specifically. The signal spectrum to be compensated is shown by the solid curve; this also represents the sweep rate of the appropriate nonlinear sweep. The effect of the nonlinear sweep is exactly inverse to this; the result is a flat signal spectrum. The effect on the noise is the square-root effect; for every 2 dB of signal compensation there is a 1-dB rise in the noise.

Since we now understand the problem, we can devise a solution. One such solution is merely the old brute-force tactic—carry on sweeping. Thus, in Fig. 11-28a we reproduce Fig. 11-26b, and then in Fig. 11-28b we see the effect of sending out more of the same nonlinear sweeps. Sooner or later, we extend the signal-to-noise bandwidth to the signal bandwidth. But this, as is usual with the brute-force approach, is inefficient; much of the extra sweeping time is spent giving an unnecessarily good signal-to-noise ratio at the middle frequencies.

Clearly, no solution can avoid the need for *some* extra sweeping time. The best solution, therefore, does not spend any extra sweeping time near those middle frequencies, but concentrates it at the edge frequencies, where the

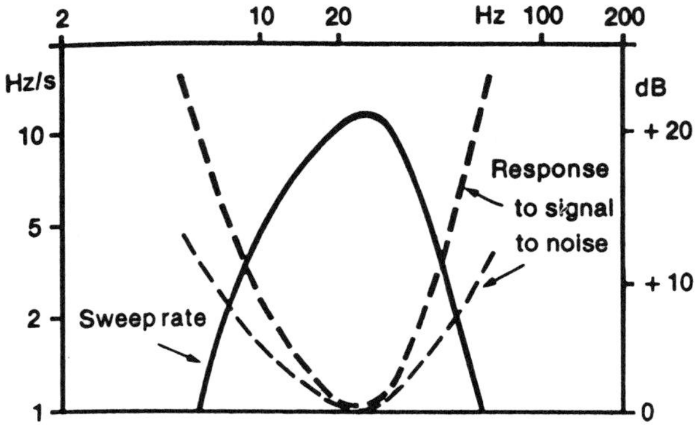

Fig. 11-27 Another way of visualizing the result of Fig. 11-26b. The nonlinear sweep specified by the sweep rate shown (solid line) affects the signal in an exactly inverse manner (thick dashes) and the noise in a manner that is the square root of this (thin dashes).

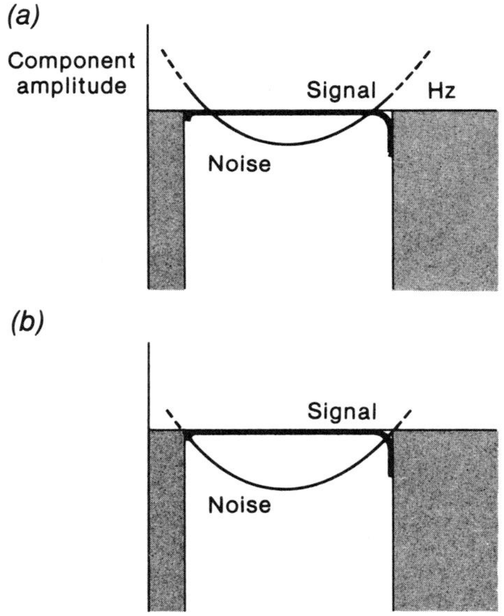

Fig. 11-28 Brute-force solution to the problem of the rising noise response.

signal-to-noise ratio is poorest. In other words, the best solution involves *overcompensation* of the signal spectrum.

Figure 11-29a shows again the signal and noise spectra obtained with a linear sweep. At midband, the signal exceeds the noise by (let us say) a safe margin w. Figure 11-29b shows again the effect of using a nonlinear sweep calculated to flatten the signal spectrum; at midband we have kept the sweep

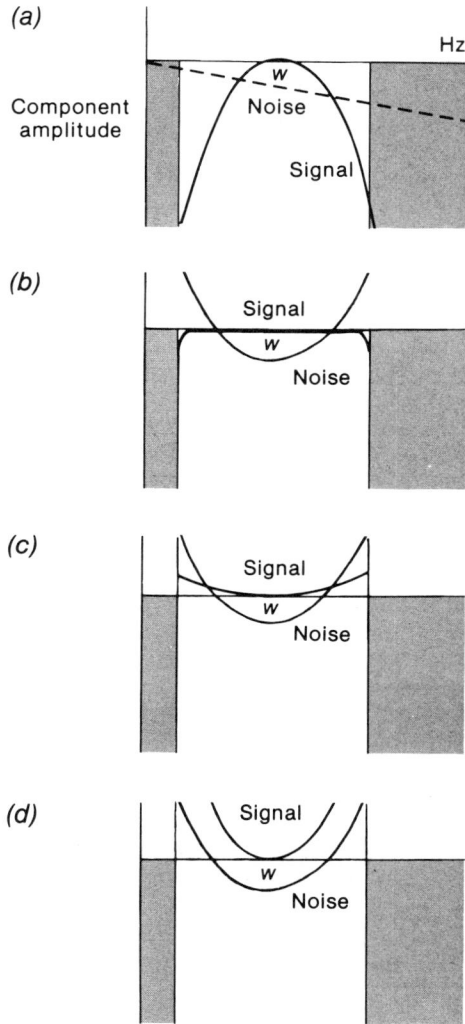

Fig. 11-29 Overcompensating the signal spectrum. (a) Signal and noise spectra for a linear sweep. (b) The same after the use of a nonlinear sweep calculated to *flatten* the signal spectrum. (c) The same after the use of a nonlinear sweep calculated to give partial *overcompensation* of the signal spectrum. (d) The same after the use of a nonlinear sweep calculated to *invert* the signal spectrum.

rate the same as for the linear sweep, and so the signal-to-noise margin remains *w*.

Figure 11-29c shows what happens as we begin to overcompensate the signal spectrum; the signal spectrum turns concave, but at the ends it is catching up to the noise spectrum. Since we still keep the sweep rate unchanged at

(a)

Sweep
amplitude

(b)

Component
amplitude

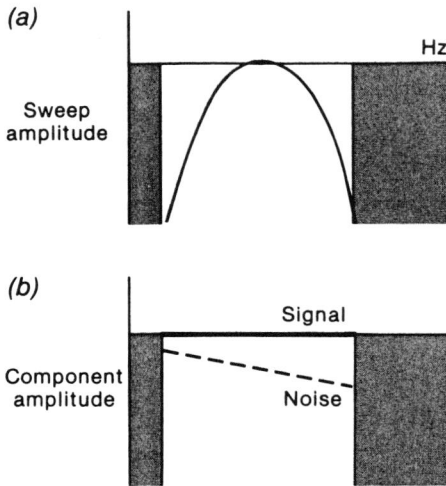

Fig. 11-30 Achieving the desired signal-to-noise spectrum. (a) Amplitude–frequency relation of the sweep used as control sweep in the correlation process. (b) Final signal and noise spectra, achieved by correlating the sweep of (a) with the signal of Fig. 11-29d.

midband, the signal-to-noise margin w is also unchanged. Figure 11-29d takes the overcompensation to the limit, at which the signal spectrum is exactly inverted. At this point it has overtaken the noise spectrum. However, we have still wasted no effort at midband; the margin w remains. Thus *this approach gives the most efficient way of enlarging the signal-to-noise bandwidth: we use a nonlinear sweep whose sweep rate is proportional to the square of the signal spectrum* (or, better yet, the square of the signal-to-noise spectrum).

This is not the end of the solution: we wish to have a flat signal spectrum. Having achieved a signal-to-noise bandwidth equal to the sweep bandwidth, we could leave this flattening to the deconvolution program. However, it is better to use a determined inverse filter for a determined effect, and the best inverse filter is again given by manipulating the *amplitude* of the control sweep against which we correlate. Therefore, *we correlate against a control sweep whose amplitude as a function of frequency follows the original signal spectrum,* as in Fig. 11-30a. The output is then given at 11-30b. We have achieved our ultimate objective—at target level the signal spectrum is flat, and it is above the noise over the full bandwidth of the sweep. And this has been done with the minimum effort; the amount of energy emitted at midband is the same as for the old linear sweep.

Unfortunately, although the effort is minimum, it is not small. We were already somewhat dismayed by the sweeping time necessary to *flatten* the signal spectrum. What will it be now, to *invert* the signal spectrum?

Figure 11-31a is the specimen vibrator response that we used before, in Fig. 11-16a. Now, however, the response required of the nonlinear sweep is inverse to the square of this; we see it at (b). Using the equations of Appendix E, we compute the corresponding variations of sweep rate and time with frequency, and obtain Figs. 11-32 and 11-33.

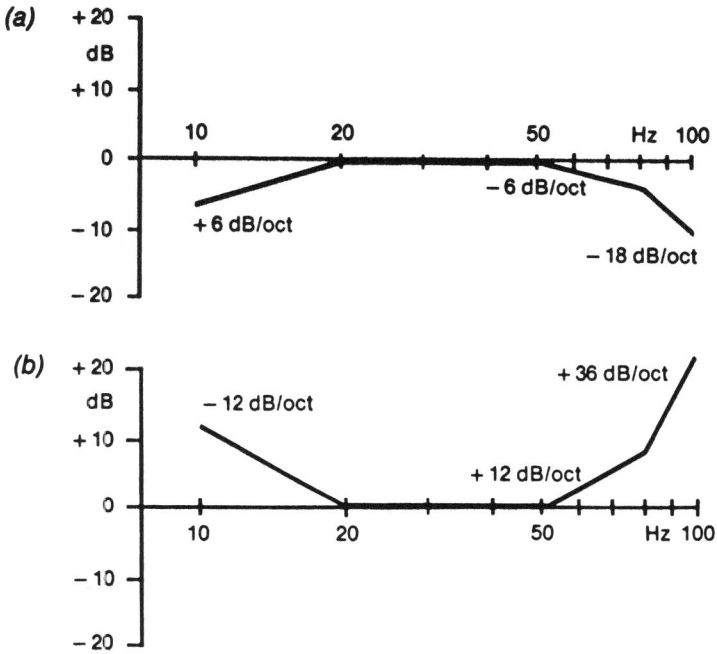

Fig. 11-31 Counterpart of Fig. 11-16 when the object of the nonlinear sweep is to invert the response of the vibrator. (a) Vibrator response itself. (b) Required characteristic of the nonlinear sweep.

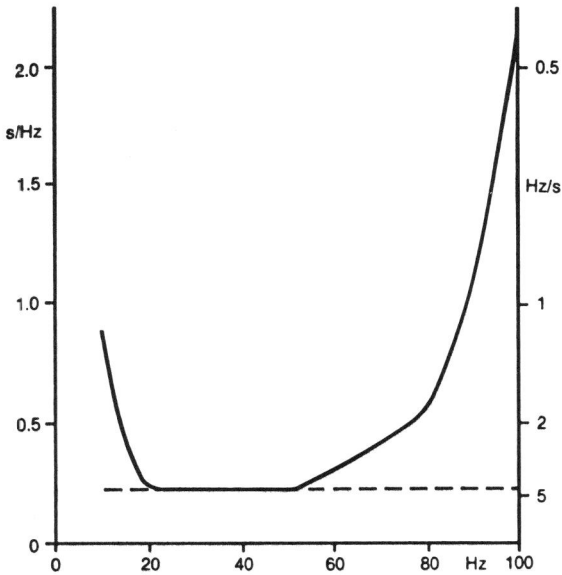

Fig. 11-32 Sweep rate (and the inverse sweep rate) of the nonlinear sweep that inverts the response of the vibrator.

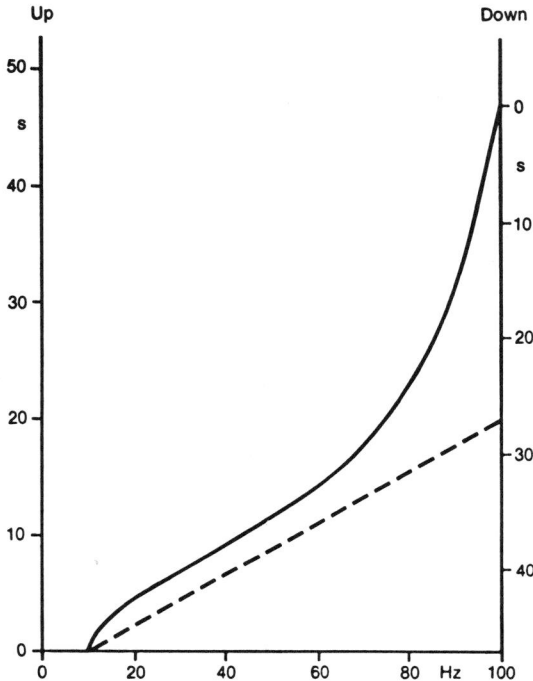

Fig. 11-33 Frequency–time relation for the nonlinear sweep that inverts the response of the vibrator.

As always, the price we pay at the low frequencies is small (2.2 s). At high frequencies, the price becomes large; Fig. 11-32 shows that we pay nearly 2.2 s for the single hertz between 99 and 100 Hz. The total duration of the sweep is now 47 s (compared to 20 s for the linear sweep and about 29 s for the signal-flattening sweep of Fig. 11-17b).

But the vibrator response is only the beginning; now we should add the earth, as we did before. Figures 11-34a and b repeat the graphs of Figs. 11-23a and b, showing the combined action of the vibrator and the earth. Figure 11-34c squares and inverts the combined response, showing what we are asking the nonlinear sweep to provide—more than 40 dB (or 100 : 1) of increase at 100 Hz. We are braced for some long sweeping times.

Using the equations of Appendix E, Fig. 11-35 shows the corresponding variation of sweep rate and inverse sweep rate. It now costs us nearly 25 s (which is more than the entire duration of the original linear sweep) to obtain the single hertz between 99 and 100 Hz. Figure 11-36 shows the sweep itself; its total duration is now—yes—184 s.

Unless our exploration objective definitely requires the resolution given by a bandwidth to 100 Hz, we probably decide to be less ambitious. Figure 11-36 contains within it the trade-off between bandwidth and cost. Relating the

(a)

(b)

(c)

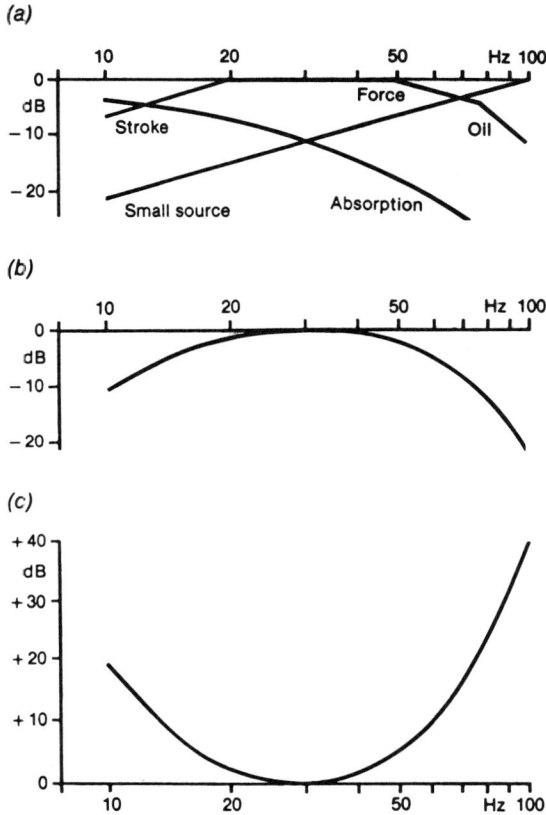

Fig. 11-34 Inverting the combined losses of Fig. 11-23. (a) Spectrum-shaping agencies. (b) Their combined effect. (c) Squared characteristic now required of the nonlinear sweep.

full line to the dashed line (that is, the nonlinear sweep to the linear sweep), we can see that twice the sweeping time gives us bandwidth to about 65 Hz, three times gives us bandwidth to about 80 Hz, and four times gives us bandwidth to about 86 Hz. Figure 11-37 graphs the relation.

The steeply rising cost of extending the bandwidth beyond 80 Hz is a consequence of the compressibility of the vibrator oil and the assumed value of 0.3 dB/Hz for the product of absorption and the target time. The cost is less daunting if we have better vibrators, or less absorption, or shallower targets.

If our choice of nonlinear sweep is exactly right for the prescribed target level, correlation against the appropriate sweep yields a target reflection whose spectrum is flat between the limits of the sweep. Shallower reflections are excessively rich in high frequencies, because they have suffered less absorption. We must therefore apply a calculated time-variant filter in processing, to bring the spectrum of these shallow reflections also to flatness.

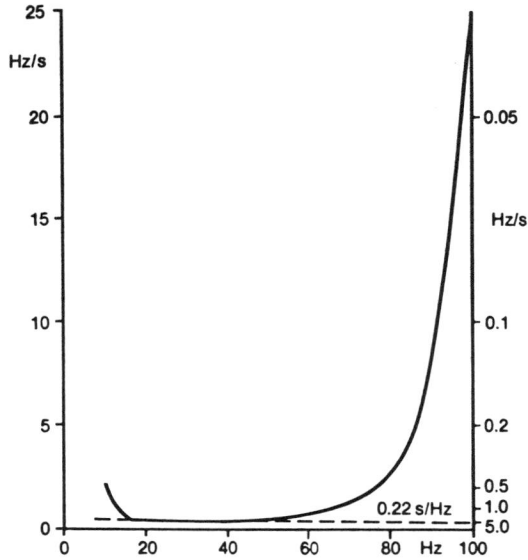

Fig. 11-35 Sweep rate (and inverse sweep rate) for the nonlinear sweep appropriate to the characteristic of Fig. 11-34c. The scaling of the vertical axis is *10 times* that used for the previous sweep-rate diagrams.

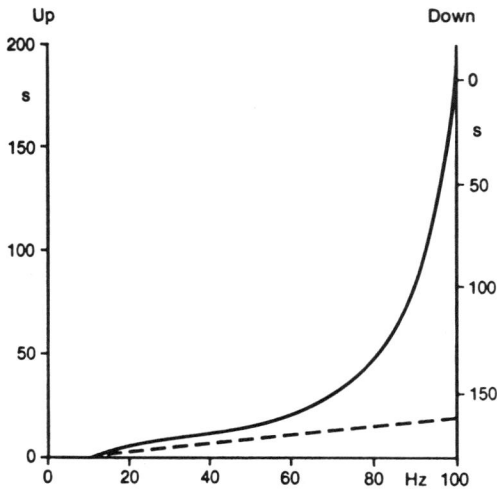

Fig. 11-36 Nonlinear sweep to invert the vibrator–earth response of Fig. 11-34b. The scaling of the time axis is *five times* that used for the previous sweep diagrams.

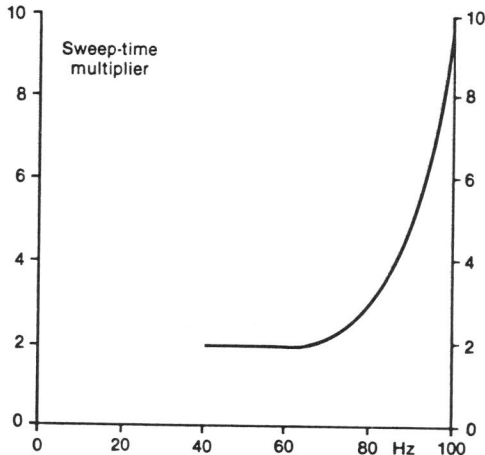

Fig. 11-37   Proportionate increase in sweeping time necessary to extend the signal-to-noise bandwidth from 10 to 40, 50, 60, . . . Hz.

11.3.7 Conclusions

Our problem is to improve on the vertical resolution given by a linear sweep. With the linear sweep, we obtain a certain reflection spectrum at target level, and a certain noise spectrum. The ratio between the two is the signal-to-noise spectrum, and the frequencies at which the signal falls below the noise define the signal-to-noise bandwidth. Then our problem is to increase the signal-to-noise bandwidth.

The most efficient way to do this is to use a nonlinear sweep whose sweep rate is proportional to the *square* of the signal-to-noise spectrum. If we have no knowledge of the noise spectrum and we assume it to be flat, then the sweep rate is proportional to the square of the signal spectrum.

As a first approximation to the signal spectrum, we may take the known response of the vibrator and a logically estimated response for the earth. The calculation of the nonlinear sweep may then be done very simply, from simple equations. An example of such a sweep is given in Fig. 11-36.

The use of a sweep rate proportional to the *square* of the signal spectrum yields a record very rich in the frequencies that were previously very attenuated. The signal spectrum is deliberately overcompensated to raise it above the noise at these frequencies. The overcompensation is then corrected by correlating against a control sweep whose *amplitude* follows the signal spectrum. The result is a flat signal spectrum at target level.

Closer approximations may be obtained if we can measure the frequency response of the vibrator–ground coupling, and the combined response of absorption, scattering, and short-path multiples at target level. The effect of array size can also be taken into account, if a restriction to small dips has allowed the arrays to be large.

Nonlinear sweeps calculated as above represent only a modest loss of production if the frequencies to which bandwidth is required are themselves modest. However, they represent a formidable loss of production if we demand bandwidth to very high frequencies. In such a case, we must do everything cost-effectively possible to minimize the noise. We must also drive the vibrator hard at the high frequencies.

When we have done everything we can to minimize the noise and to improve the vibrator signal, the nonlinear sweep calculated as above is the most efficient way to achieve the desired signal-to-noise bandwidth. The cost may be high, but we can be assured that *there is no better way.*

CHAPTER 12

The High-Resolution
Vibroseis Survey
In Practice

From all the foregoing, we see how important it is to *design the field work to be self-consistent.* If we just do what we were doing before, but merely say, "Let's try a nonlinear sweep, and see if we get better resolution," the result will certainly not be as good as it should be, and will probably be worse than before. We shall be left saying, "We tried those nonlinear sweeps, but they don't work here."

Which is nonsense. *Nonlinear sweeps work everywhere,* provided we apply them as part of a total and harmonious scheme. Let us construct that scheme for a situation where a high-resolution detail survey is required over a prospect found by a previous reconnaissance survey.

The first task, as always, is to understand the exploration problem. We go and talk to the interpreters. They are amazed and delighted; no one from the field ever asked them such questions. We look over the reconnaissance survey with them until we understand the nature of the exploration play and its seismic expression. We ask to what degree the prospect hinges on the delineation of the faults, and whether the faults are nearly vertical at levels of interest. We study the sections together, and read the labels, to see whether the reconnaissance techniques are likely to have obscured steep dips and faults. We study the maps together, to see the orientation of the reconnaissance lines relative to the structural grain, the faults, and any important stratigraphic feature. From the maps we read off the maximum dips in milliseconds per kilometer or per mile. We ask whether there is a single target, on which we can concentrate exclusively, or multiple targets. We establish the range of travel times for the target, and the reflection velocity estimated in the reconnaissance processing; we also establish an average dominant "period" of the reflections in the target zone. We ask particularly what thickness of reservoir can be expected, and whether we can hope for a gas–liquid contact within the reservoir.

If there is a well in the area, we review the seismic sections across it in comparison to those across the new prospect. Provided that the well is indeed relevant, we learn everything we can about reservoir thickness, interval velocity, and even porosity. If we can actually follow the reservoir from the well to the new prospect, we look on the sections for any evidence of changes of reservoir thickness or geologic setting. And then we construct synthetic seismograms, with a range of pulse bandwidths, to show the way in which the solution of the exploration problem depends on the reflection bandwidth.

In some cases, then, we emerge with a clear resolution requirement—a bandwidth requirement—that *must* be satisfied if the detail survey is to be a success. In particular, we determine f_m, the highest necessary reflection frequency. Then our problem is to satisfy that requirement. Of course, this should be at minimum cost, but we recognize that the minimum cost may be high. The constraint is that the cost of exploration by seismics must be less than the cost of exploration by drilling alone.

In other cases, as we agreed once before, the requirement may be merely for "better" resolution; then our problem is to estimate the trade-off between cost and resolution, and to make a sensible judgment. We shall follow these two situations through the following material.

We take our leave of the interpreters and go to the archives. We are interested in looking at the processing tests (particularly any with larger bandwidth than that finally adopted) and at the field tests (particularly those where the number and duration of the sweeps were decided). However, there is one measurement of paramount importance we need to make on the old data, and for that we shall probably need to do some reprocessing.

The measurement is to determine *what sweeping time was necessary to raise the midband signal above the noise at target level.* We remember our basic diagram, in Fig. 12-1a, with the signal above the noise throughout the sweep. From our measurements on the reconnaissance sections we can estimate the peak of the spectrum; perhaps the apparent "period" was 33 ms, allowing us (for present purposes) to take the peak of the spectrum at 30 Hz. We are going to apply a fairly tight filter, perhaps 25 to 35 Hz, and determine from the reconnaissance survey what sweeping time was necessary to bring the signal in *this* band above the noise.

It is unlikely, but there is just a chance that we can find the field tapes of the individual sweeps before vertical stacking. Then we can *reprocess a short stacked section over a representative piece of line, using one, two, four, and eight sweeps per source point.* Through the narrow-band filter we see the target

(a)

Component
amplitude

14   46   Hz

(b)

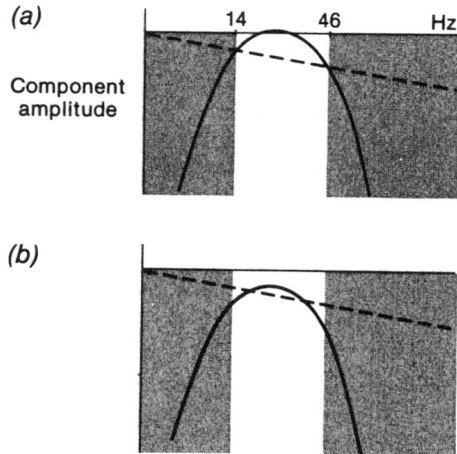

Fig. 12-1 Reminder of the likely situation from the reconnaissance survey, at target level. (a) With all the sweeps used. (b) With one-quarter of the sweeps used.

reflection beginning to poke its nose above the noise with, perhaps, two sweeps (Fig. 12-1b).

More likely, we find little or no test material preserved, and nothing but vertically stacked field tapes. Then we do the reprocessing with *a range of fold in the common-midpoint (cmp) stack,* from single-fold to full fold. Again we look for the fold that, through the narrow-band filter, just makes the target reflection evident above the noise. Perhaps this is six-fold, in a situation where full coverage was 24-fold. Then, to a first approximation, we can say that one-quarter of the sweeps per source point, coupled with 24-fold cmp stack, would have done the same.

We are not saying here that the reconnaissance survey used more sweeps than it should have. More sweeps were necessary, both to get a bandwidth greater than 25 to 35 Hz (by the brute-force approach of Fig. 12-1a) and to get reflections deeper than the present target. For the new survey, we shall endeavor to get increased bandwidth at target level by more efficient means; we also accept, of course, that *we shall not do as well as the reconnaissance survey on reflections below the target.*

Let us suppose that the reconnaissance survey used eight linear sweeps of 14 to 46 Hz, each 7 s long. The sweep rate was about 4.6 Hz/s and the time per hertz about 0.22 s. Since we found that one-quarter of these sweeps (that is, two) would have been sufficient to see the signal at the peak of the spectrum, we now know that it takes about 0.44 s/Hz to do this.

As soon as we have this figure established, we can start to plan the new survey. We ask about crews in the area; we find one with four vibrators, each rated at 105 kN (24,150 lb-force). This is good; the reconnaissance survey was done with three vibrators rated at 100 kN (23,000 lb-force). Since the signal-to-noise ratio varies directly with the number and output of the vibrators and with the square root of the sweeping time, we now need less than 0.44 s/Hz at the peak of the signal spectrum; in fact we need only 0.44 (^{3}/$_4$ × ^{100}/$_{105}$)^2, or 0.22 s/Hz. One of the original sweeps, with the four new vibrators, would have done as well as two with the three old vibrators.

We now have the initial design requirement for our nonlinear sweep. The sweep (or all the sweeps together, if we have to use more than one) must have an inverse sweep rate of 0.22 s/Hz at 30 Hz.

We ask the vibrator manufacturer at what frequencies we can expect the low- and high-frequency droops in the vibrator response. He replies: 6 dB/octave below 20 Hz, − 6 dB/octave between 50 and 80 Hz, and − 18 dB/octave above 80 Hz. It sounds familiar; we remember Fig. 11-34a, repeated here as Fig. 12-2a. The same.

Then—the earth. There being no one we can ask about this, we make a first guess. We know that we shall be using small arrays, and that therefore we

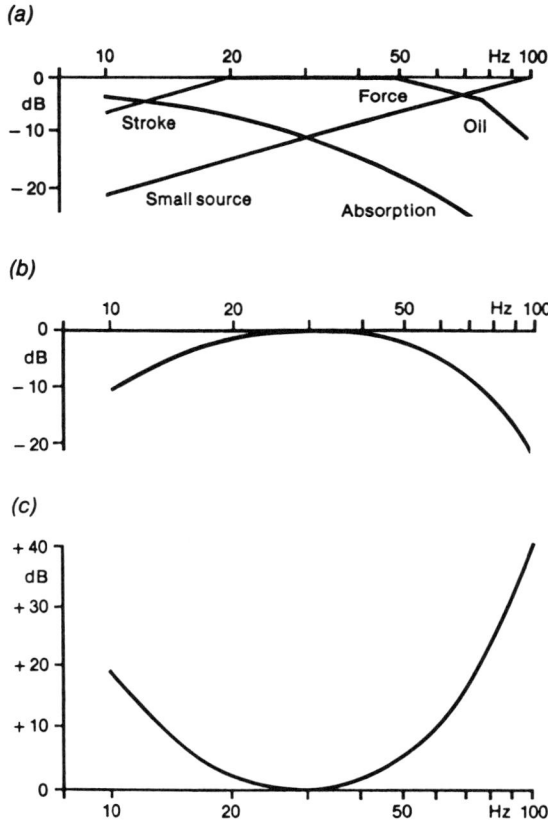

Fig. 12-2' Reminder of the combined spectrum-shaping agencies and the requirements of the nonlinear sweep.

can expect the small-source response in Fig. 12-2a to extend across the whole spectrum. In general, we cannot determine from the reconnaissance survey what high-frequency attenuation the earth will impose, because the reconnaissance sweep did not extend to sufficiently high frequencies. But we do know the travel time to the target; let us say it is 1.5 s. Our first guess at the absorption might be 0.2 dB/cycle; then the overall absorption loss becomes 0.3 dB/Hz. Just like Fig. 12-2a again.

So our first guess at the nonlinear sweep has the squared response of Fig. 12-2c. It must also have, we said, an inverse sweep rate of 0.22 s/Hz at its reference frequency of 30 Hz.

We go through the calculations, and find (oddly enough) that this sweep is exactly that of Figs. 11-35 and 11-36, repeated here as Figs. 12-3 and 12-4. The maximum inverse sweep rate in Fig. 12-3 is indeed 0.22 s/Hz, at 30 Hz. So, in Fig. 12-4, we have our first estimate of the nonlinear sweep to use for the high resolution survey.

Next, we must plan the spread geometry. This we do using the principles set out in Chapter 10. We remember that all we need is the target depth, the reflection velocity, the design dip, and the highest frequency. The first three we know from the reconnaissance survey.

For a velocity of 3000 m/s (10,000 ft/s), for example, we can graph the group interval as a function of the highest frequency, with design dip as a parameter (Fig. 12-5a). The group interval, in turn, defines the source interval.

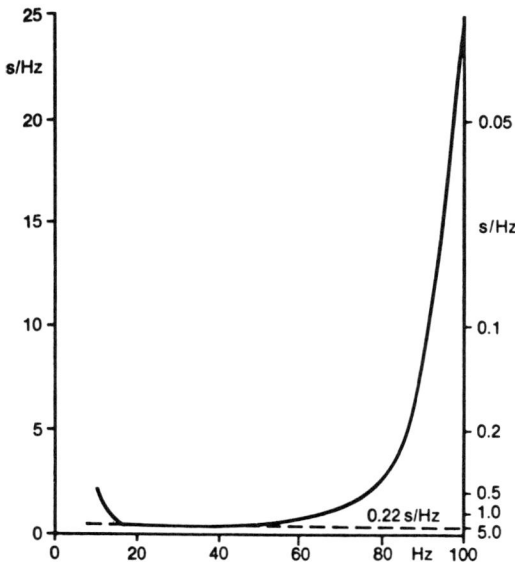

Fig. 12-3 Variation of sweep rate (and inverse sweep rate) for the nonlinear sweep appropriate to the characteristic of Fig. 12-2.

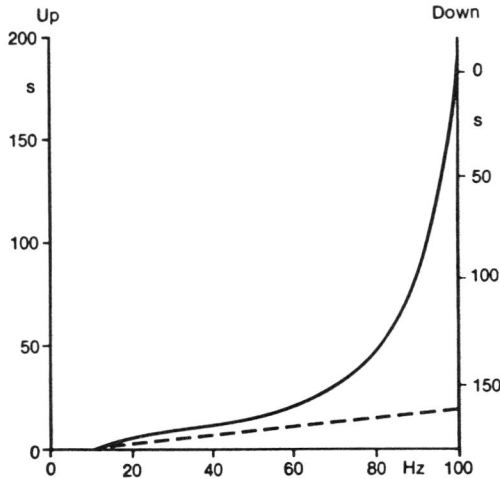

Fig. 12-4 Nonlinear sweep to invert the vibrator–earth response of Fig. 12-2b.

To proceed further, we must consider the practical limitations on the sweep. Perhaps the maximum length of signal that the correlator can handle is 32 s. We probably adopt a cycle time of 31 s; a maximum reflection time of 6 s then sets the maximum sweep duration at 25 s. From Fig. 12-4 we can see that, if our highest-frequency objective is about 65 Hz, one 25-s sweep is sufficient. But if we wish for 70 Hz, we must split into two 15-s sweeps, which then need two 21-s cycles. An objective of 100 Hz requires ten 19-s sweeps, and so ten 25-s cycles.

From these considerations we can graph the minimum time per kilometer (or per mile) as a function of the highest frequency selected; this we do, for the same range of design dips, in Fig. 12-5b. The figures allow a reasonable 20 s for the vibrators to move between source points. Also shown in Fig. 12-5b is the time taken by the original reconnaissance survey, which used eight 7-s sweeps and a source interval of 70 m (220 ft).

Now we make the economic decisions. At one extreme, the exploration problem may demand a highest frequency of 100 Hz. We can see from Fig. 12-5 that if the design dip is 18° the group interval is 45 m (150 ft), and the time is 100 min/km or 160 min/mile. Without allowance for time lost (cable delays, breakdowns, angry farmers, and so on), this is only one-quarter of the production of the reconnaissance crew. If the design dip is 90°, the time rises to about 5 hours (h)/km or 8 h/mile.

At the other extreme, we may ask merely for the highest frequency we can get for a defined cost—perhaps 50% more than that of the reconnaissance survey. Then we see that we can achieve a highest frequency of 90 Hz if the design dip is 18°, 83 Hz if the design dip is 30°, and 68 Hz if it is 90°.

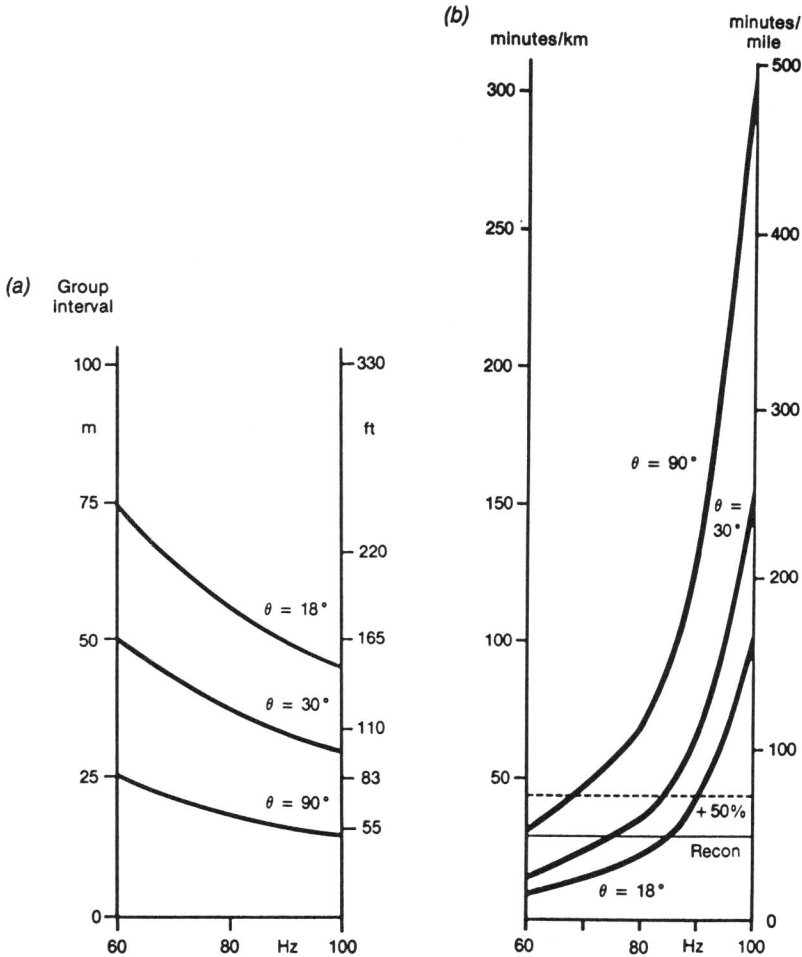

Fig. 12-5   How the judgment of maximum frequency affects production. (a) Variation of group interval with maximum frequency for three choices of dip. (b) Variation of production with maximum frequency for three choices of dip. The maximum sweep length is taken as 25 s; the maximum reflection time is 6 s; and the time to move between source points is 20 s.

Since the design dip is established from the reconnaissance survey, Fig. 12-5 effectively confronts us with our fundamental decision: *resolution costs money—how much are we prepared to pay?*

When we have answered this question in light of the exploration problem, we can read off the group interval, order our spread cables, burn the nonlinear sweep into the memory chips, and go to the field. We have done everything we can do by thought; now the crew is on the ground, and the meter is running.

The first day we probably go to one of the old reconnaissance lines and repeat the sweeping regime of the reconnaissance survey (eight 7-s sweeps at 14 to 46 Hz) over each new source interval. We record the sweeps individually, without vertical stacking. We have seen that the starting point of the new sweep design is the sweep time necessary to get the signal above the noise at the peak of the signal spectrum. We knew this result from the old data, but it is worth taking a day to confirm it, and to establish a standard of comparison for the later nonlinear operations. If we have processing facilities close at hand (and this is highly desirable for high-resolution work), we try to process the first day's work before the next morning. Again, our prime concern is to establish the sweep time per hertz necessary to raise the signal above the noise, at target level, over the narrow band of 25 to 35 Hz.

If the results confirm the earlier figure (or if we have to wait for the results), we restart the line on the second day, using our previously calculated nonlinear sweep. We remember to do everything we reasonably can to minimize the noise. We also do everything we can to preserve the smooth *regularity* of the field operations. A high-resolution survey is normally smoother than a reconnaissance survey, because a greater proportion of the time is spent sweeping; we need to take all possible advantage of this.

If the results from the first day differ significantly from our expectations, the variable we change is the *number* of sweeps. If the signal-to-noise ratio at the peak of the spectrum is poorer than we hoped (perhaps because our arrays are now shorter), we must increase the number of sweeps. This may lead us to review the economics; in consequence, we may be led to reduce the maximum frequency and to increase the group interval. But we do not change the *type* of nonlinearity; we are still following Fig. 12-4. If the signal-to-noise ratio at the peak of the spectrum is better than we expected (perhaps because of our care to minimize the noise or because the vibrators are in better condition), we reduce the number of sweeps. If the number is only one, we can reduce its duration. Again we are led to review the economics; perhaps we can now afford to increase the maximum frequency and to reduce the group interval.

With these decisions made and implemented, we continue for long enough to obtain a meaningful extent of processed section. Then we perform a frequency analysis on a short (200 to 300 ms) segment of processed trace that includes the target, and average this over many stacked traces. Now we are checking the *type* of nonlinearity: did we make a good estimate of the frequency-selective actions of the vibrator and the earth? If the reflection spectrum is substantially flat (apart from the inevitable notches, of course), then our sweep design was correct. If it is not, we made some poor estimates; perhaps the vibrator–ground coupling was sharply peaked, or short-path multiples imposed a major high-frequency loss. So we take a smoothed version of the reflection spectrum, use it to modify the response curve of Fig. 12-2b, and compute a new nonlinear sweep. At this stage of refinement, we can also introduce compensation for any decrease in low-frequency drive, which is desirable to prevent vibrator decoupling, and for any increase in high-fre-quency drive we are able to achieve without overheating the vibrator.

We remember not to attack a poor reflection *spectrum* with more sweeps. It works, but we cannot afford the inefficiency of brute force. The right approach is always to modify the nonlinearity.

We also remember that we appraise a high-resolution section differently from a reconnaissance section. The criterion is resolution at target level—not overall prettiness or ease of picking. Furthermore, we must think carefully about what to expect *below* the target level. A reconnaissance section is nor-mally filtered progressively lower at greater depth; this has the effect of giving penetration, at progressively lower frequencies. The nonlinear sweep designed to optimize the resolution at target level gives a quite different appearance; it preserves the full bandwidth of the sweep at the target, with acceptable signal-to-noise ratio, but it gives virtually *nothing* below the target (unless the reflec-tions have significantly larger reflection coefficients, of course). *All frequencies drop below the noise together.* Consequently, nothing can be done to extend the penetration by time-variant filtering.

For some exploration problems, this may not be acceptable. Perhaps we need evidence of a path for hydrocarbon migration below the reservoir, or the direction of a fault below the target. Then we have two choices. The first, and more expensive, is to set the design target depth below the actual reservoir depth. The second is to design for a wide, flat spectrum at the actual target depth, as we have done, but to supplement the nonlinear sweep at the low and middle frequencies. This, we remember, costs us little. However, it does mean that the deeper reflections are then present, at the lower frequencies, and that the application of some time-variant high-cut filtering *below* the target is likely to recover them.

Just a few small points left, and then we are done.

Nowhere in this section have we expressed any preference between up-sweeps and downsweeps. In Chapter 8, we remember, we discussed their relative advantages and disadvantages. The downsweep locks the phase-com-pensation circuitry more quickly, compacts the surface more quickly, and keeps the accumulators full through most of the sweep. The upsweep avoids the risk of serious harmonic ghosts. On balance, then, the downsweep is preferable when-ever the sweep bandwidth is so small or the sweep duration is so long that harmonic ghosts cannot arise. With modern long sweeps, this often means that we prefer downsweeps. When we change to nonlinear sweeps, we have to reconsider this.

Perhaps we are able to avoid the generation of harmonics caused by baseplate decoupling at the low frequencies; this we do, we remember, by reducing the drive level at the low frequencies, and then compensating the loss by decreasing the sweep rate at these frequencies. This might allow us to maintain a preference for downsweeps. Otherwise, we must check at what frequencies the harmonics are generated, and be sure that the sweep rate at and above those frequencies is not such as to generate harmonic ghosts within the duration of the record.

For example, the danger point in the sweep represented in Fig. 12-3 is at the low-frequency end of the flat portion of the curve, where the sweep rate is 4.5 Hz/s; if distortion is severe at 18 Hz, a harmonic ghost at 36 Hz will appear at 4 s after every event. This may be unacceptable; if no other change is easy, we would revert to upsweeps.

We must be particularly conscious of this problem where we are striving for the very high frequencies, and therefore using many sweeps. Then the appropriate sweep rate to use in the harmonic ghost calculation is the sweep rate on the *individual* sweeps.

How about diversity stacking? We have agreed several times that resolution problems lead us to use the longest sweeps we can. If our resolution objectives are very demanding, we still use many of these sweeps, and so we can be completely comfortable using the diversity-stacking technique of Chapter 10. This technique relies on having several or many sweeps, vertically stacked, at each source point. If our resolution objectives are less demanding, we have seen from Fig. 12-5 that we may need only one or two long sweeps. In this case we must review the need for diversity stacking, and find a balance; we are trading the noise advantage of diversity stacking against the loss of efficiency in splitting one or two long sweeps into six or eight shorter ones.

How about processing? It is very important that high-resolution field work be followed by high-resolution processing. After all that expenditure of thought and money in the field, it would be a catastrophe to throw away all the benefits by poorly considered processing. The processors must understand what we have done and why we have done it. The first case in point arises if we are using the nonlinear sweeps to invert the signal spectrum, but not doing the correlation in the field; then we must be sure that the processors know that *the correlation must be performed against a sweep whose amplitude is modulated to represent the signal spectrum,* in accord with Fig. 11-30a.

Another case in point is that spatial data reduction (for example, adding groups in pairs) is unlikely to be permissible. Another concerns deconvolution; deterministic inverse filtering (particularly for the response of the geophones and the low-cut instrumental filter) may be preferable before stack, but cheaper after stack. Statistical deconvolution should not be done before the signal-to-noise benefit of the stack, *nor before the time-variant filter that flattens the signal spectrum above the target.* Deconvolution gates should not extend into the region of poor signal-to-noise ratio below the target. And, of course, no filtering is ever permissible at target level. (The only possible exception arises if we have indeed achieved a flat and rectangular signal spectrum at target level; then, right at the end, after migration, we may apply a little rounding to the shoulders of the spectrum in order to reduce the distant side lobes of the reflection pulse.)

Figure 12-6 illustrates the real benefit obtainable with nonlinear sweeps.

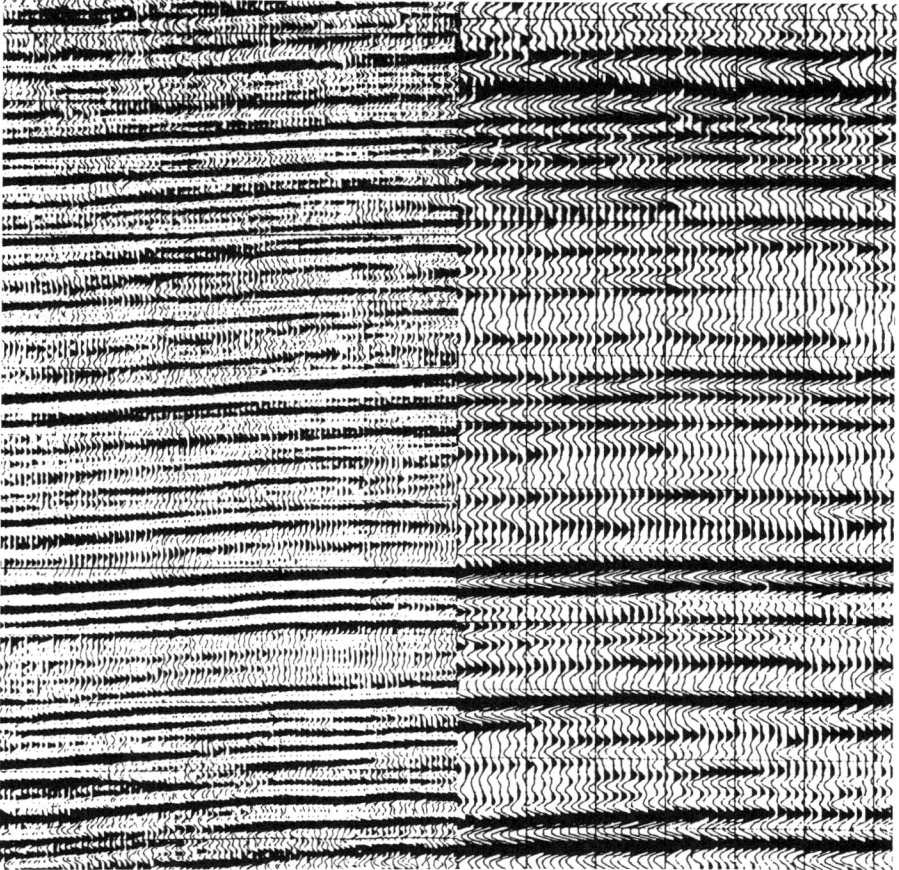

Fig. 12-6   Difference between traditional techniques (right) and techniques chosen for high resolution (left) (courtesy of Strat-Seis).

When we first consider the profusion of variables in the Vibroseis field technique, and the complex interrelationships among them, we are bewildered. How can we possibly choose? When we add the variables in the processing, the situation is worse. When we add an infinite selection of nonlinear sweeps, the situation appears hopeless.

We find little help in the literature. Individual papers and articles address individual choices, but cause conflicts when we have to balance one desired objective against another exclusive one. As a result, the choice of variables in many field crews has been incoherent—even contradictory.

What an unexpected pleasure it is, then, to find that the selection of field and processing techniques can be made rational, straightforward—even easy. And that this includes nonlinear sweeps.

We remember what we must decide first. We must establish a target time (which may be that of basement, if exploration interest is present at all levels) and a target velocity. We must establish a maximum dip, which may be 90° if we wish the maximum possible resolution of vertical faults. We must establish how long the reconnaissance sweeps dwelt at the peak signal frequency, to get an acceptable signal-to-noise ratio there; this is easily determined by a little inexpensive reprocessing of the reconnaissance survey. We must know the manufacturer's specifications for the vibrators and the geophones, and we must make a first estimate of the absorption in the earth.

Thereafter, everything follows. In particular, we can construct a graph (Fig. 12-5b) showing the trade-off between cost and resolution. We make our judgment, and the other major variables are determined. Then we can refine the sweep, if we wish, to optimize the cost-effectiveness.

With this approach, we can be confident that the technique is *harmonious within itself*. What an unexpected pleasure.

Some Details

1. The linear sweep is given by

$$s(t) = A \sin 2\pi \left[f_1 + \frac{(f_2 - f_1)\, t}{2T} \right] t$$

where f_1 is start frequency, f_2 is end frequency, A is amplitude, and T is duration. (We note the factor 2 in the denominator of the last term. The basic equation is $A \sin \theta$, and the phase angle is the time integral of frequency; the factor 2 is a consequence of this integration.)

2. The cross-correlation function of two series $a(t)$ and $b(t)$ is

$$\phi_{ab}(\tau) = \int_{-\infty}^{\infty} a(t) \cdot b(t + \tau)\, dt$$

where τ is the shift between the two series (so that the variation of τ slides one past the other).

3. The auto-correlation function of the series $a(t)$ is therefore

$$\phi_{aa}(\tau) = \int_{-\infty}^{\infty} a(t) \cdot a(t + \tau)\, dt$$

4. The auto-correlation function of the sweep (to a sufficient approximation) is

$$\phi_{ss}(\tau) = \frac{A^2 T}{2} \frac{\sin \pi (f_2 - f_1)\tau}{\pi(f_2 - f_1)\tau} \cos 2\pi \left\{ \frac{f_2 + f_1}{2} + \frac{(f_2 - f_1)\tau}{2T} \right\} \tau$$

The first factor is the zero-shift (or zero-lag) value, which represents the energy in the sweep. The second factor is the sinc function that defines the envelope. In the third factor, the term $(f_2 - f_1)\tau/2T$ can be neglected for sweeps and auto-correlations of usual length; in this case the third factor is simply the mid-frequency carrier, whose amplitude is then modulated by the sinc function. The ratio of the first trough of the auto-correlation to the central peak is the square root of the ratio of the end frequencies of the sweep.

5. The cross-correlation function of two series $a(t)$ and $b(t)$ is the same as the convolution of the two series in which either series is reversed in time. Thus the result of cross-correlating $a(t)$ against $b(t)$ is the same as the result of passing $a(t)$ through a filter whose impulse response is $b(-t)$, or of passing $b(t)$ through a filter whose impulse response is $a(-t)$. The operation of cross-correlating the geophone signal $g(t)$ against the control sweep $s(t)$ is therefore equivalent to passing the geophone signal through a filter whose impulse response is the sweep reversed in time. This filter is the **matched filter** for the sweep. If the amplitude spectrum of the sweep is taken as rectangular, this filter has an amplitude-frequency response of the same rectangular form, and a phase-frequency response that exactly cancels that of the reflected sweeps contained in the geophone signal.

6. The amplitude spectrum of a constant-amplitude linear sweep is not exactly rectangular, but may be taken as so for most purposes. The abrupt termination of the sweep at the end frequencies leads to a superabundance of the end frequencies in the spectrum; the details are given in Goupillaud (1976).

7. If the amplitude of a linear sweep is not constant, its amplitude spectrum, and hence the amplitude-frequency response of the correlation, approximately follows the amplitude variations of the sweep. Thus tapering of the ends of a sweep represents a filter whose response can be visualized from the envelope of the sweep. If (as in much past practice) this tapering is applied *both* to the sweep sent to the vibrators *and* to the sweep sent to the correlator, this filtering is applied twice; the amplitude-frequency response is the *square* of the tapering. This is pointlessly harsh, and the practice should be discontinued. Ideally, no tapering for side-lobe reduction should be applied until after the deconvolution process.

8. Side lobes at distant times are sometimes called "correlation noise." As noted in the main text, these are not a problem in standard Vibroseis practice; we can see that they are there (at very low level) where the gain is high preceding the first breaks, but they do not obtrude in the body of the record. In part, however, this is because we deliberately adopt field techniques that minimize them. Normally, for example, the ground roll forces us to use a long offset and, traditionally, long arrays. Both of these have the effect of reducing the amplitude of the first breaks relative to reflections in the body of the record. If we are working in an area untroubled by ground roll, and take advantage of the opportunity to reduce the offset and the array lengths, we increase the amplitude of the first breaks. This may cause the distant side lobes of the high-amplitude first breaks to be larger than the reflections in the body of the record. (Our test, obviously, would be to squint across the record at the angle of the first breaks, searching for events parallel to them.) Even if this were not so on the monitor record, it might become so after deconvolution; the deconvolution acts to reduce the tapering effect of the earth filter, to sharpen the edges of the spectrum at the limits of the sweep, and so to increase the correlation noise. For operation with short offsets and short arrays, therefore, various schemes exist to reduce the risk of damage from correlation noise. The three basic principles are:

- To use sequences of *short sweeps* (so that the correlation noise is necessarily zero beyond the duration of the sweep)
- To use sweeps of slightly *different sweep rates* over each source array, so that the correlation noise is reduced by vertical stacking; this solution, obviously, requires correlation before vertical stacking, and therefore needs a fast correlator
- To use special signals selected to have minimum (or even zero) correlation noise; these are the **pseudorandom** signals

Details of these three approaches may be found in Cunningham (1979) and Edelmann and Werner (1982).

9. The earth filter acts only on the reflected sweeps, and so is applied only once (as in explosive work). Any desired filter may be applied to the sweep used in correlation, by modulating the amplitude or phase of that sweep. Because the sweep (the operator) is so long, arbitrary responses can be realized that would be expensive or unstable to apply in any other fashion. A very sharp notch filter of zero phase is a case in point.

10. We develop this notch-filter example for the case where the field equipment provides the option of such a notch, and where power-line pickup (say at 60 Hz) is extreme. Can the problem be dealt with in later processing, or should the observer use analog notch filters in the field? The observer makes a recording of the same SP with and without the analog filters and then correlates each against a sweep in which the 59- to 61-Hz portion is zeroed. If the penetration is better on the record with analog filters, the observer should use them. If not, the pickup is not so serious that it is driving the signal too far below the system noise. A correlation-type notch filter in the processing can handle the problem. This is more desirable than an analog filter, because the latter has a 180° phase shift across the notch. Both filters, of course, make the reflections ringy at the frequencies each side of the notch.

11. Implicit in the above paragraphs is the fact that the amplitude spectrum of the auto-correlation function is the square of the amplitude spectrum (that is, the power spectrum) of the sweep. If a linear sweep has a constant amplitude, the amplitude spectrum of the sweep and the auto-correlation function are both rectangular. This is the reason, of course, why the elementary discussion of Fig. 2-3 (in terms of a phase response only) yields a pulse shape the same as the auto-correlation of Fig. 3-3. It is also the reason why we would not choose to correlate the geophone signal against the actual signal emitted into the earth by an (imperfect) vibrator.

12. The mixed-phase nature of the Vibroseis wavelet is unwelcome, because in the deconvolution process we prefer that our wavelets be minimum phase. The solution is easy: during the correlation or the early processing, we apply an operator having the minimum-phase response approximately corresponding to the sweep spectrum. Then, since the result of convolving two minimum-phase responses (the operator and the earth) is itself minimum phase, the Vibroseis wavelet can be taken as minimum phase, and deconvolution can proceed on this basis.

13. Our initial discussion of Vibroseis was in the Fourier terms of Fig. 2-9; the white box has an amplitude response that is the rectangular amplitude spectrum of the sweep, and a phase response (a system of frequency-dependent delays) that is the negative of the phase spectrum of the sweep. This view was introduced first because it is more physically satisfying than correlation. It is easier to see that all noise outside the sweep spectrum is excluded, and that all the energy strung out in the sweep is compressed into a short pulse. With this concept established, the discussion then turned to correlation, because correlation is the historical method of achieving the required response in practice. Today, however, much so-called correlation is actually performed in the frequency domain, because present hardware makes this the cheapest method. As and if this becomes universal, it will be possible to describe Vibroseis completely without ever mentioning the word correlation.

14. Let us suppose that we place a geophone down a deep borehole, and a single vibrator at perhaps 100 m from the top of the hole. The vibrator emits a sine wave; we record the output of the geophone in the steady state and measure its amplitude. Now we bring up a second (identical) vibrator, side by side with the first, and drive them both with the same sine wave. What happens to the downhole amplitude?

Clearly, we have increased the mechanical power by a factor of 2. Our first expectation, therefore, is that the seismic power would also increase by a factor of 2; this would mean that the downhole amplitude would increase by a factor of $\sqrt{2}$.

In fact, recent controlled experiments show that the downhole amplitude increases by a factor of 2. We are doubling the mechanical power, but obtaining four times the seismic power. At frequencies for which the source is small relative to a wavelength, this is because a larger source is more efficient in converting mechanical power to seismic power. We can assume that the reflected wave shows the same relationships. So, in terms of signal-to-noise ratio, which is the better: to emit one sweep from two vibrators side-by-side or to emit two sweeps from one vibrator and stack them?

Suppose that, instead of bringing the second vibrator side-by-side with the first, we place it 100 m from the top of the hole on the opposite side. The vibrators are now 200 m apart, and the borehole geophone is equidistant from them. Again we drive them with the same sine wave. What happens to the downhole amplitude? It doubles, but for a different reason. There is no longer any significant increase in the efficiency of the vibrators; the downhole amplitude doubles by simple superposition. But there is no contradiction of conservation of energy; the radiation is now directional, and where we get more than a twofold increase in one direction, we get less than a twofold increase in another. In the downward direction (or in any other direction for which the phase difference between the two radiated sine waves is an even multiple of π), the amplitude doubles, approximately. But for any direction in which the phase difference is an odd multiple of π, the amplitude is zero, approximately. Averaged over all directions, the seismic power is increased by a factor of 2, but no more.

And so we come to the crunch question and an important practical issue. We space the vibrators apart by a distance (perhaps 30 m) not large enough to produce destructive interference of the compressional wave in any direction, but large enough for us to be unsure whether this is still a simple point source. What happens to the downhole amplitude? Does it increase by $\sqrt{2}$, or by 2, or by something in between?

15. We are in charge of three seismic crews. Crew A is shooting dynamite, crew B is using Vibroseis for a deep target, and crew C is using Vibroseis for a shallow target. The recording equipment on all three crews is very old. We have enough money in the budget to refit one crew with new equipment, whose main characteristic (apart from considerations of reliability and convenience) is lower input noise and higher signal level for the same distortion. There is just a possibility we may get the money to refit two crews. The party manager on the dynamite crew complains a great deal. Which equipment would we replace first, and which second?

APPENDIX B

The Decibel-per-Octave Sweep

The equation for a sweep whose auto-correlation function has a spectrum rising in proportion to frequency is

$$f = \left[\left(f_2^2 - f_1^2 \right) \frac{t}{T} + f_1^2 \right]^{1/2}$$

where f is the instantaneous frequency at time t, f_1 and f_2 are the start and end frequencies, and T is the duration of the sweep.

For a spectrum rising with the square of frequency, the equation is

$$f = \left[\left(f_2^3 - f_1^3 \right) \frac{t}{T} + f_1^3 \right]^{1/3}$$

In general, for a spectrum varying with the nth power of frequency, the equation is

$$f = \left[\left(f_2^{n+1} - f_1^{n+1} \right) \frac{t}{T} + f_1^{n+1} \right]^{1/n+1}, \quad n \neq -1$$

(For purposes of hand calculation, the computational failure of this equation at $n = -1$ can be overcome by using a value of n very close to -1.)

The sweep rate as a function of time is obtained merely by differentiation; however, the sweep rate as a function of frequency is both simpler and more useful:

$$\frac{df}{dt} = \frac{f_2^{n+1} - f_1^{n+1}}{(n+1)\,Tf^n}$$

From this we can see directly that the sweep rate is inversely proportional (and the spectrum therefore directly proportional) to the nth power of frequency.

If we rearrange the general equation above to give t as a function of f, we see that we could call this type of sweep a **power-law** sweep.

APPENDIX C

The Decibel-per-Hertz Sweep

As before, it is more instructive to treat frequency as the independent variable. If the absorption in the earth is a decibels per cycle, and the travel time to the target and back is t_0, the absorption at frequency f is $at_0 f$ decibels. The frequency response of the earth path is proportional to $e^{-0.115 a t_0 f}$, where the factor 0.115 represents the conversion from decibels to natural logarithms. The sweep rate must be inverse to this. The resulting equation for the sweep, using b to represent $0.115 a t_0$, is

$$t = T \frac{e^{bf} - e^{bf_1}}{e^{bf_2} - e^{bf_1}}.$$

The sweep rate is

$$\frac{df}{dt} = \frac{e^{bf_2} - e^{bf_1}}{Tb} e^{-bf}$$

When frequency is taken as the independent variable, this sweep is naturally called an exponential sweep. If we insist on using time as the independent variable, the sweep appears as a logarithmic sweep.

APPENDIX D

Calculation of Nonlinear Sweeps for Cascaded Responses

We can practice the calculation of nonlinear sweeps on the example of Fig. 11-23 (repeated here as Fig. D-1). We remember the five agencies affecting the combined response of the vibrator and the earth, as illustrated in Fig. D-1a. The three agencies concerned with the vibrator operate only in restricted frequency bands. The absorption of the earth operates across the entire band. The small-source effect operates below a corner frequency determined by the size of the source array; to allow good bandwidth at steep dips, the figure assumes a corner frequency at the upper limit of the sweep.

The overall response of Fig. D-1b at each frequency is obtained by the multiplication (decibel addition) of the relevant individual responses. The response of the vibrator below 20 Hz is given by $f/20$. From 50 to 80 Hz it is $50/f$, and from 80 to 100 Hz there is an additional factor of $80^2/f^2$. The response of the small source is $f/100$. The response of the absorptive earth is e^{-bf}, where $b = 0.0345$.

Multiplying together the relevant responses, we have that the overall response is defined as follows:

$$10\text{--}20 \text{ Hz}, \qquad \frac{f^2 e^{-bf}}{2000}$$

$$20\text{--}50 \text{ Hz}, \qquad \frac{f e^{-bf}}{100}$$

(a)

(b)

(c)

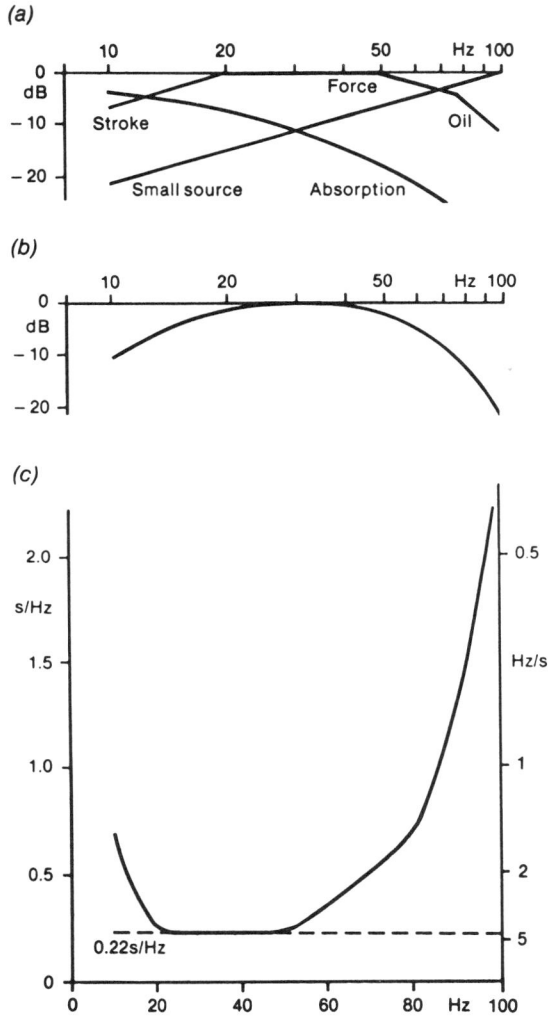

Fig. D-1   Reprise of Fig. 11-23.

$$50\text{–}80 \text{ Hz,} \qquad \frac{e^{-bf}}{2}$$

$$80\text{–}100 \text{ Hz,} \qquad \frac{3200e^{-bf}}{f^2}$$

To obtain the overall response of Fig. D-1b, we must normalize the response to a 0-dB value at its peak. We can see by eye that the peak occurs between 20 and 50 Hz, so we differentiate the expression $fe^{-bf}/100$ and set this to zero to find the peak. It is at 29 Hz, where the response is 0.107 (-19.4 dB).

The responses relative to this peak are therefore obtained by dividing the above expressions by 0.107. These are the values plotted in Fig. D-1b.

The next stage is to calculate the inverse sweep rate as a function of frequency. In Fig. D-1c, we take this as inverse to the above responses, and set the value at the peak of the response ($f = 29$ Hz) to be the same as that of the reference linear sweep. The reference linear sweep, we remember, is 10 to 100 Hz in 20 s, for which the inverse sweep rate is $20/(100 - 10)$, or 0.222 s/Hz. Therefore, the inverse sweep rate for 20 to 50 Hz (for example) becomes $100 \times 0.107 \times 0.222/fe^{-bf}$. Doing the arithmetic, we have

$$10\text{--}20 \text{ Hz}, \qquad \frac{47.5e^{bf}}{f^2} \quad \frac{\text{s}}{\text{Hz}}$$

$$20\text{--}50 \text{ Hz}, \qquad \frac{2.37e^{bf}}{f} \quad \frac{\text{s}}{\text{Hz}}$$

$$50\text{--}80 \text{ Hz}, \qquad 0.0474e^{bf} \quad \frac{\text{s}}{\text{Hz}}$$

$$80\text{--}100 \text{ Hz}, \qquad 7.42 \times 10^{-6} f^2 e^{bf} \quad \frac{\text{s}}{\text{Hz}}$$

The frequency–time relations of the four parts of the sweep are then obtained by integrating these expressions.

$$10\text{--}20 \text{ Hz}: \quad t = 47.4\left[b \left(\ln f + bf + \frac{b^2 f^2}{2 \times 2!} + \frac{b^3 f^3}{3 \times 3!} + \cdots \right) - \frac{e^{bf}}{f} + 0.0488 \right]$$

$$20\text{--}50 \text{ Hz}: \quad t = 2.37\left(\ln f + bf + \frac{b^2 f^2}{2 \times 2!} + \frac{b^3 f^3}{3 \times 3!} + \cdots -3.83 \right)$$

$$50\text{--}80 \text{ Hz}: \quad t = 1.37e^{bf} - 7.71$$

$$80\text{--}100 \text{ Hz}: \quad t = 0.215\left[f^2 e^{bf} - 2(bf - 1)\frac{e^{bf}}{b^2} + 27{,}200 \right]$$

Evaluating the time of each sweep at its highest frequency, we have

10–20 Hz,	$T =$	3.843 s
20–50 Hz,	$T =$	7.028 s
50–80 Hz,	$T =$	13.944 s
80–100 Hz,	$T =$	28.103 s

Thus the total time of the sweep is the sum of these, or 52.968 s.

APPENDIX E

Calculation
of Nonlinear Sweeps
to Invert
Cascaded Responses

The calculation starts with the same four responses as we used in Appendix D. These are then squared. The steps of normalization to the peak, inversion, and setting to the reference sweep rate are then done as before. For the example worked in Appendix D, the results are

$$10\text{--}20 \text{ Hz}, \qquad \frac{10107 \ e^{2bf}}{f^4} \quad \frac{\text{s}}{\text{Hz}}$$

$$20\text{--}50 \text{ Hz}, \qquad \frac{25.267 \ e^{2bf}}{f^2} \quad \frac{\text{s}}{\text{Hz}}$$

$$50\text{--}80 \text{ Hz}, \qquad 0.01011 \ e^{2bf} \quad \frac{\text{s}}{\text{Hz}}$$

$$80\text{--}100 \text{ Hz}, \qquad 2.4675 \times 10^{-10} f^4 e^{2bf} \quad \frac{\text{s}}{\text{Hz}}$$

The frequency–time relations are again obtained by integration, using the general equations

$$\int f^n \ e^{2bf} df = \frac{f^n \ e^{2bf}}{2b} - \frac{n}{2b} \int f^{n-1} \ e^{2bf} \ df$$

and

$$\int \frac{e^{2bf}}{f} \ df = \ln f + 2bf + \frac{(2bf)^2}{2 \times 2!} + \frac{(2bf)^3}{3 \times 3!} + \cdots$$

The final frequency–time relations contain many terms, but are quite straightforward; they may be evaluated with a scientific pocket calculator. The constant of integration is determined by setting $t = 0$ at the start frequency of each part of the sweep.

FURTHER READING

Conveniently, many of the papers in this list have now been assembled and reprinted in a single volume: Geyer, R. L. (ed.), 1989, Vibroseis; No. 11, Geophysics Reprint Series, Society of Exploration Geophysicists, Tulsa, OK.

ANSTEY, N. A., 1966, Correlation techniques—a review, *Journal of the Canadian SEG* 2-1.

————, 1983, *Introduction to Field Work;* Manual GP301 of the Video Library for E & P Specialists, IHRDC, Boston.

————, 1986a, *Wiggles—a graphical introduction to signal theory;* Manual GP201 of the Video Library for E & P Specialists, IHRDC, Boston.

————, 1986b, Whatever happened to ground roll? and Field techniques for high resolution. *The Leading Edge,* March, p. 40, and April, p. 26.

ARNOLD, M. E., 1977, Beam forming with vibrator arrays, *Geophysics* 42-07-1321.

BAETEN, G. J., FOKKEMA, J. T., and ZIOLKOWSKI, A. M., 1988, Seismic vibrator modelling, *Geophysical Prospecting* 36-1-22.

BERNHARDT, T., and PEACOCK, J. H., 1978, Encoding techniques for the Vibroseis system, *Geophysical Prospecting* 26-1-184.

BICKEL, S. H., 1982, The effects of noise on minimum-phase Vibroseis deconvolution, *Geophysics* 47-8-1174.

BROWN, G. L., and MOXLEY, S. D., 1964, Design considerations for vibrators, *IEEE International Convention Record* 12-8.

CHAPERON, A., and LABADEMS, B., 1983, Use of nonlinear sweeps for generation of high-frequency data in Vibroseis; paper presented at the 53rd Annual International SEG Meeting, Las Vegas, Nevada.

CHAPMAN, W. L., BROWN, G. L. and FAIR, D. W., 1981, The Vibroseis system, a high-frequency tool, *Geophysics* 46-12-1657.

CLEMENT, A. H., 1963, Correlation by random time reference utilization; U.S. Patent No. 3,108,249.

CORUH, C., and COSTAIN, J. K., 1983, Noise attenuation by Vibroseis whitening processing, *Geophysics* 48-5-543.

CRAWFORD, J. M., DOTY, W. E. N., and LEE, M. R., 1960, Continuous signal seismograph, *Geophysics* 25-1-95.

CUNNINGHAM, A. B., 1979, Some alternate vibrator signals, *Geophysics* 44-12-1901.

DANKBAAR, J. W., 1983, The wavefield generated by two vertical vibrators in phase and in counterphase, *Geophysical Prospecting* 31-2-873.

DENHAM, L. R., 1981a, Extending the resolution of seismic reflection exploration, *Journal of the CSEG* 17-1-43.

———, 1981b, Field-technique design for seismic reflection exploration; paper presented at the 51st Annual International SEG Meeting, New Orleans, La.

DOTY, W. E., AND CRAWFORD, J. M., 1954, Method and apparatus for determining travel time of signals; U. S. Patent No. 2,688,124.

DOUGHERTY, S. M., and JUSTICE, J. H., 1988, An approximation technique for determining optimal Combisweep parameters, *Geophysics* 53-7-989.

DRAGOSET, W. H., 1988, Marine vibrators and the Doppler effect, *Geophysics* 53-11-1388.

EDELMANN, H. A. K., 1966, New filtering methods with Vibroseis, *Geophysical Prospecting* 14-4-455.

———, 1982, A contribution to the investigation of amplitude characteristics of vibrator signals, *Geophysical Prospecting* 30-6-786.

———, and WERNER, H., 1982, The encoded sweep technique for Vibroseis, *Geophysics* 47-5-809.

———, KIRCHHEIMER, F., and SCHULTE, L., 1989, Decomposed vibrator patterns to improve seismic survey results, *Geophysical Prospecting* 37-2-167.

ERLINGHAGEN, L., 1976, Vibroseis: new results under different geophysical aspects; Prakla-Seismos, Hannover, Federal Republic of Germany.

FARRELL, W. E., 1978, A linear analysis of the interaction between a vibrator and an elastic medium; paper presented to San Francisco convention of SEG.

GARRIOTT, C., 1980, Digital Vibroseis velocity surveys; SSC-Birdwell, Tulsa, Okla.

GEYER, R. L., 1970, The Vibroseis system of seismic mapping, *Journal of the Canadian SEG* 6-1-39.

———, 1973, Vibroseis refraction weathering techniques, *Geophysics* 38-2-285.

———, and ANSTEY, N. A., 1985, *Array design;* Manual GP305 of the Video Library for E & P Specialists, IHRDC, Boston, Mass.

GOUPILLAUD, P. L., 1976, Signal design in the Vibroseis technique, *Geophysics* 41-6-1291.

GURBUZ, B. M., 1972, Signal enhancement of vibratory-source data in the presence of attenuation, *Geophysical Prospecting* 20-2-421.

————, 1982, Upsweep signals with high-frequency attenuation and their use in the construction of Vibroseis synthetic seismograms, *Geophysical Prospecting* 30-4-432.

HARGROVE, K. L., BONACORSI, D. G., and ANDREW, A. J., 1983, Use and misuse of the nonlinear Vibroseis method for the acquisition of high-resolution seismic data; paper presented at the 53rd Annual International SEG Meeting, Las Vegas, Nev.

KALLWEIT, R. S., and WOOD, L. C., 1982, The limits of resolution of zero-phase wavelets, *Geophysics* 47-7-1035.

KIRK, P., 1981, Vibroseis processing, in *Developments in Geophysical Exploration Methods* 2, ed. A. A. Fitch, Applied Science Publishers, London.

KOEFOED, O., 1981, Aspects of vertical seismic resolution, *Geophysical Prospecting* 29-1-21.

KREY, T., 1969, Remarks on the signal-to-noise ratio in the Vibroseis system, *Geophysical Prospecting* 17-3-206.

LANDRUM, R. A., 1980, Quality control displays for seismic vibrators; paper presented to the SEG convention in Houston, Tex.

LERWILL, W. E., 1981, The amplitude and phase response of a seismic vibrator, *Geophysical Prospecting* 29-4-1657 (see also comments on this paper by Sallas, J. J., and Weber, R. M., *Geophysical Prospecting* 30-6-935, and further discussion by Mitchell, K. L., and Edelmann, H. A. K., in *Geophysical Prospecting* 31-5-843 and 31-6-995).

LINES, L. R., and CLAYTON, R. W., 1977, A new approach to Vibroseis deconvolution, *Geophysical Prospecting* 25-3-417.

————, CLAYTON, R. W., and ULLRYCH, T. J., 1980, Impulse response models for noisy Vibroseis data, *Geophysical Prospecting* 28-1-49.

MILLER, G. F., and PURSEY, H., 1954, The field and radiation impedance of mechanical radiators on the free surface of a semi-infinite isotropic solid, *Proceedings of the Royal Society* A 223–521.

————, and ————, 1956, On the partition of energy between elastic waves in a semi-infinite solid, *Proceedings of the Royal Society* A 225–55.

MORSE, P. F., and HILDEBRANDT, G. F., 1989, Ground-roll suppression by the stack array, *Geophysics* 54-3-290.

OKAYA, D. A., and JARCHOW, C. M., 1989, Extraction of deep crustal reflections from shallow Vibroseis data using extended correlation, *Geophysics* 54-5-555.

PRITCHETT, W.C., 1990, *Acquiring Better Seismic Data,* Chapman and Hall, London.

RIETSCH, E., 1977a, Computerized analysis of Vibroseis signal similarity, *Geophysical Prospecting* 25-3-541.

————, 1977b, Vibroseis signals with prescribed power spectrum, *Geophysical Prospecting* 25-4-613.

————, 1981, Reduction of harmonic distortion in vibratory source records, *Geophysical Prospecting* 29-2-178.

RISTOW, D., and JURCZYK, D., 1975, Vibroseis deconvolution, *Geophysical Prospecting* 23-2-363.

SAFAR, M. H., 1984, On the determination of the downgoing P-waves radiated by the vertical seismic vibrator, *Geophysical Prospecting* 32-2-392.

SALLAS, J. J., 1984, Seismic vibrator control and the downgoing P-wave, *Geophysics* 49-6-732.

SCHRODT, J. K., 1987, Technique for improving Vibroseis data, *Geophysics* 52-4-469.

SERIF, A. J., and KIM, W. H., 1970, The effect of harmonic distortion in the use of vibratory surface sources, *Geophysics* 35-2-234.

SHERIFF, R. E., 1989, *Geophysical Methods,* Prentice-Hall, Englewood Cliffs, N. J.

SODBINOW, E. S., 1985, *Multiple Coverage;* Manual GP304 of the Video Library for E & P Specialists, IHRDC, Boston.

TAN, T. H., 1985, The elastodynamic field of N interacting vibrators, *Geophysics* 50-8-1229.

THIGPEN, B. B., DALBY, A. E., and LANDRUM, R., 1975, Special Report of the Subcommittee on Polarity Standards, *Geophysics* 40-4-694.

WATERS, K. H., 1978, *Reflection Seismology,* Wiley, New York.

WIDESS, M. B., 1973, How thin is a thin bed? *Geophysics* 38-5-1176.

INDEX